MARCH FOR SCIENCE

EDITED BY Stephanie Fine Sasse
and Lucky Tran, PhD

The MIT Press
Cambridge, Massachusetts
London, England

Designed by Sloane Henningsen and Stephanie Fine Sasse

© 2018 Stephanie Fine Sasse and Lucky Tran

This book was set in Lato, Merriweather, and Oswald by the design team. Printed and bound in Canada.

Library of Congress Cataloging-in-Publication Data is available.

ISBN:

978-0-262-03810-2

10 9 8 7 6 5 4 3 2 1

Some may consider science the purview of a special or separate type of citizen, one who pursues natural facts and generates numerical models for their own sakes. But our numbers here today show the world that science is for all. Our lawmakers must know and accept that science serves every one of us. Every citizen of every nation and society. Science must shape policy. Science is universal. Science brings out the best in us. **With an informed optimistic view of the future, together we can — dare I say it — save the world.**

— Bill Nye
Honorary Co-Chair, March for Science

Science seeks to answer "why" by seeking facts. Facts that can be tested and verified. We believe that evidence must be reproducible. We believe in the power of doubt. We are comfortable with uncertainty. Science touches all aspects of modern life. Fundamental basic science has made possible the cars we drive, the houses we live in, the clothes we wear, our smartphones, our entertainment, the weapons that keep us safe. Basic science underlies the medical advances that allow us to lead longer, healthier lives… **Support our future. Invest in science.**

— Lydia Villa-Komaroff, PhD
Honorary Co-Chair, March for Science

It is time for all of us to fight back against those who deny science and those who degrade science. It is time for all of us to step out of our clinics, our classrooms, and our labs. We need to make ourselves known into the halls of our government. **We need to hear all of your voices.**

— Mona Hanna-Attisha, MD
Honorary Co-Chair, March for Science

To everyone who marched and keeps marching — in spirit or in person. To every volunteer who donated their time, skills, voice, or support. To the activists who came before us and made this movement possible and the future activists who will sustain it.

Thank you. This book is for you.

I MARCH BECAUSE I WANT TO BE A SCIENTIST WHEN I GROW UP. I MARCH SO MY SON CAN GROW UP HEALTHY. I MARCH TO HOLD MY REPRESENTATIVES ACCOUNTABLE. I MARCH TO ENSURE THAT SCIENCE IS OPENLY COMMUNICATED. I MARCH TO SUPPORT THE EPA. I MARCH SO THAT I CAN LIVE TO BE 100. I MARCH BECAUSE I'M A TREKKIE SO I SUPPORT NASA. I MARCH BECAUSE I WANT A FUTURE. I MARCH FOR A DIVERSE SCIENCE THAT SERVES EVERYBODY. I MARCH TO PROTECT INDIGENOUS LANDS. I MARCH BECAUSE MY FAVORITE THINGS ARE MADE OF SCIENCE. I MARCH BECAUSE I OWE SCIENCE MY LIFE. I MARCH TO RAISE AWARENESS OF ENDANGERED SPECIES AND WHAT WE CAN DO TO SAVE THEM. I MARCH TO PROMOTE REASONABLE POLICIES INFORMED BY EVIDENCE. I MARCH BECAUSE SCIENCE GAVE ME AWESOME ROBOT LEGS TO RUN WITH. I MARCH FOR TURTLES. I MARCH FOR FOOD SCIENCE. I MARCH BECAUSE I'M AFRAID OF WHAT HAPPENS IF WE START CONFUSING STRONGLY HELD OPINIONS WITH FACTS. I MARCH BECAUSE I NEED A CURE FOR MY GRANDDAUGHTER'S CANCER. I MARCH BECAUSE CLIMATE CHANGE IS NOT A CONSPIRACY. I MARCH BECAUSE SCIENCE IS POLITICAL, BUT IT SHOULDN'T BE PARTISAN. I MARCH TO PROTECT THE PAST, PRESENT, AND FUTURE. I MARCH BECAUSE ANTHROPOLOGY AND SOCIOLOGY ARE OUR BEST WAYS OF UNDERSTANDING SOCIETY AND WE SHOULD USE THEM. I MARCH TO PROTECT THE OUTDOORS THAT I LOVE TO EXPLORE. I MARCH TO SUPPORT MARGINALIZED COMMUNITIES. I MARCH FOR ANSWERS. I MARCH TO INCREASE FUNDING FOR SCIENTISTS. I MARCH FOR THE NIH. I MARCH FOR THE PLANET. I MARCH FOR BETTER HEALTHCARE. I MARCH BECAUSE MY HOUSE WAS DESTROYED IN A HURRICANE AND SCIENCE HELPED ME ESCAPE IN TIME TO SAVE MY FAMILY. I MARCH BECAUSE THERE ARE NO SECOND CHANCES

WHY WE MARCH

WHEN IT COMES TO SAVING THE
BELONGS IN POLICY. I MARCH
YELLOWSTONE SUPER VOLCANO AND
TELL ME WHEN TO EXPECT IT. I MARCH
SUBJECT IN SCHOOL. I MARCH BECAUSE
BECAUSE CURIOSITY IS A VIRTUE.
MARCH FOR MY MOM WHO WENT
ME. I MARCH BECAUSE SCIENCE IS
OF MY HEROES WERE SCIENTISTS. I
MY LIFE. I MARCH BECAUSE THERE
MARCH BECAUSE I WILL NEVER HAVE
OF MY GRANDPARENTS' GENERATION.
SAVED MY PUP FROM PARVO. I MARCH
MY SISTER GAVE ME A KIDNEY TO
POSSIBLE. I MARCH BECAUSE SCIENCE

PLANET. I MARCH BECAUSE SCIENCE
BECAUSE I AM TERRIFIED OF THE
I'D REALLY PREFER FOR SCIENCE TO
BECAUSE SCIENCE IS MY FAVORITE
I LIKE TO MAKE THINGS. I MARCH
I MARCH FOR NATIONAL PARKS. I
BACK TO GET HER PHD AFTER RAISING
AWESOME. I MARCH BECAUSE MOST
MARCH BECAUSE SCIENCE CHANGED
IS LITERALLY NO REASON NOT TO. I
TO DEAL WITH THE TERRIBLE DISEASES
I MARCH FOR THE TREATMENT THAT
FOR PROGRESS. I MARCH BECAUSE
SAVE MY LIFE, BUT SCIENCE MADE IT
IS THE FOUNDATION OF ALL OF THE

TECHNOLOGY THAT I LOVE. I MARCH BECAUSE I BELIEVE IN WHAT SCIENCE COULD BE. I MARCH BECAUSE I MET A SCIENTIST WHEN I WAS A KID WHO TOLD ME THAT THEY ASK "WHY" FOR A LIVING AND THAT SOUNDS LIKE SOMETHING WE SHOULD SUPPORT. I MARCH TO CHANGE THE STEREOTYPES ABOUT SCIENTISTS. I MARCH TO FIGHT CLIMATE CHANGE. I MARCH BECAUSE SCIENCE IS FOR EVERYONE. I MARCH FOR THE SOCIAL SCIENCES. I MARCH BECAUSE I'M COLOR BLIND AND SCIENCE MADE IT POSSIBLE FOR ME TO WEAR GLASSES THAT LET ME SEE THE DIFFERENCE BETWEEN RED AND GREEN AND IT'S BEAUTIFUL. I MARCH BECAUSE I WOULD PERSONALLY LOVE TO GO TO MARS ONE DAY. I MARCH BECAUSE SCIENCE MAKES BEER BETTER. WINE TOO ACTUALLY. I MARCH TO HOLD SCIENCE ACCOUNTABLE. I MARCH FOR MY DAUGHTER, SO SHE KNOWS PEOPLE ARE FIGHTING TO PROTECT THE EARTH FOR HER FUTURE. I MARCH BECAUSE SCIENCE MADE CRIME SHOWS EVEN BETTER. I MARCH BECAUSE YOU HAVE TO SUPPORT ANYTHING THAT FIGURED OUT HOW TO PUT ACTUAL HUMAN BEINGS ON THE MOON. I MARCH TO CELEBRATE ALL OF THE WOMEN AND PEOPLE OF COLOR WHO HAVE CONTRIBUTED TO SCIENCE THROUGHOUT HISTORY. I MARCH TO BE PART OF THE CONVERSATION. I MARCH BECAUSE WITHOUT A WARNING FROM NOAA, A TORNADO WOULD HAVE KILLED ME. I MARCH BECAUSE DINOSAURS ARE THE COOLEST THING EVER. I MARCH FOR MUSEUMS AND TEACHERS. I MARCH BECAUSE SCIENCE MADE MY BABY POSSIBLE. I MARCH BECAUSE I HAVE A JOB BECAUSE OF SCIENCE. I MARCH BECAUSE I WANT TO LIVE IN A WORLD WHERE MARCHES FOR SCIENCE AREN'T NECESSARY. I MARCH FOR A MORE INCLUSIVE SCIENTIFIC COMMUNITY. I MARCH SO THAT WE CAN ALL HAVE SUPERHUMAN POWERS ONE DAY. I MARCH FOR _____.

INTRODUCTION

On the evening of April 22 — and for many weeks after — volunteers and supporters around the world gathered to debrief the largest event in support of science in history. Our conversations did not revolve around the number of people who showed up in each city or the number of tweets or shares on social media. Instead, we talked about the signs, the speeches, the chants, and the stories that moved us — the voices of the March for Science movement.

Everyone who participated in the March for Science had their own personal reason for marching: science saved our life or the life of someone we love, gave us new opportunities, challenged us, inspired us, or compelled us to work toward a better world. Many of us overcame barriers to be part of the scientific community and marched so that the next generation might not have to struggle the way we did. Others marched in honor of our heroes, our doctors, our teachers, or our family. Some of us marched with our children, each step taken toward the future they deserve. Wherever we marched, and whatever we marched for, we marched together, connected by the hope that science in service of the common good can change the world.

More than one million people (and at least five penguins) joined together across all seven continents to send a message. We took to the streets — or, in some cases, to the sea — to remind our governments, our politicians, and our peers that science belongs to everyone, and that a threat to science is a threat to all of us. We led chants that demanded a central role for evidence in policymaking. We carried signs that promoted the advancement of scientific research that improves our health, strengthens our economy, powers technological innovation, ignites the curiosity of individuals and communities, and contributes to our collective understanding of the world around us. Science has long been a tool for building a healthy, imaginative, and thriving society — and now society is banding together to protect it.

We made our voices heard in over 600 locations, from Washington, DC, to Berlin, Germany, from the North Pole to Lusaka, Zambia, from Mt. Everest to the ocean floor. Some led marches of one, others joined crowds greater than one hundred thousand. Researchers shared their discoveries and questions with their community. Educators taught us new ways of experimenting and thinking critically about our world. Artists encouraged us to explore multiple ways of knowing. Activists helped us craft the movement's future. Across cities and towns, we wrote letters protesting cuts in funding and demanding respect for scientific consensus in policy.

This book includes a small sample of stories from the thousands who made the March for Science movement possible — the people who organized marches around the world, the communities and organizations who have worked tirelessly for years to build the foundation that we stand on and with, the supporters who marched on April 22, and the advocates who continue to march for science every day. It is about our refusal to stand by while science is at risk of censorship, while evidence is confused with opinion, or while the success of critical scientific agencies is threatened by negligence or lack of resources. It is about the ways that our lives are entangled with the innovations, discoveries, and protections that science can offer. This book is about what science was, is, and can be. Each story is a call to action to uphold the role of science in our policies, our classrooms, and our neighborhoods.

The March for Science represents the commitment of the people to inquiry, to reason, to innovation, and to growth. It catalyzed a powerful force that continues to this day. In our streets, in our homes, in our elections, and in everyday conversation, we keep marching. Until there is evidence-based decision-making and legislative processes, we keep marching. Until there is equitable science education, we keep marching. Until members of marginalized communities are on equal footing in science and equally reap its benefits, we keep marching. Until fair policies, meaningful outreach, and accountability are standard both within and surrounding scientific institutions, we keep marching. And until we are certain that we live in a society committed to protecting and improving our lives, our planet, and our future, we keep marching.

Our work is far from over. March for Science will continue to support and defend the role of science in society and policy as a collective voice demanding:

Science, not silence.

We knew something was wrong with our water. It was brown and it smelt weird and tasted gross. It would burn my skin and give me and my family rashes.

Over 8,000 kids under age six were exposed to lead. Listen to me: when we don't believe in science, and especially when our government doesn't believe in science, kids get hurt. That's what happened in Flint. **For the sake of Flint kids and for kids all over this world, I march for science.**

- Amariyanna "Little Miss Flint" Copeny
Flint, MI, USA

RALEIGH, NC, USA

ATHENS, OH, USA

I DON'T COME FROM A PEDIGREED FAMILY OF SCIENTISTS,

and I'm the first one in memory to go for a PhD in anything. But I do come from a line of strong people who understand hard work and sacrifice, of farmers and caregivers, of medical doctors and soldiers, and even a political prisoner and politician, who was assassinated the year after I was born.

I was raised by my Lola (which means grandmother in Tagalog) Polly and Ate (sister — but used as a sign of respect for any older woman) Gina in Manila, Philippines and surrounded by two dozen cousins and aunts and uncles in a family "compound." My earliest memories include saying goodbye to my mother when I was three years old, standing on a couch so I could put my arms around her neck for a hug before she flew off to work in Bahrain. We wouldn't reunite for years. My dad had already left a couple years prior and was living in the US; I didn't meet him until I was six.

When we moved from Manila to Ohio, I remember there was some turbulence as we were flying over Hawaii, and I got sick in the plane. I don't recall the nausea, only that my heart felt tied to home and my Lola, had felt stretched and taut like a tripwire. I was thinking, where were we headed really? Why did we leave a place of comfort toward something unknown? Those are questions I still ask myself, but they evolved from being asked in childlike fear to being embraced in pursuit of remarkable journeys as an adult.

My circuitous path from that six-year-old getting sick on a flight to three decades later has been full of discovery. And discovery is something science is exceedingly good at doing, which is one reason I gravitated toward it. My interest in science has given me experiences like smelling thousand-year-old air as I excavated ancient stone buildings in thick

jungles so we can document and learn about our past, and uncovering story after story with each layered removal of dirt. (Next time, I should tell you the story about waking up to a giant spider literally on my face!)

That interest led to me being invited to join a global team of scientists and scholars, spanning fields from Linguistics to Geologic Climatology, working together on a shared problem and communicating it in the highest-impact science publications in the world. It gave me the opportunity to talk about the role of climate change in the ancient Maya world for a video that was screened at the 2014 United Nations Convention on Climate Change.

More recently, it took me to the frontline of organizing the largest globally coordinated event for science advocacy — both making history and setting a personal record for weight loss.

But as I said, I come from a long line of workers and providers, and I know there is so much more to be done in service to science and society. Thankfully, one of the many, many things archaeology has taught me about people is that we write our history, and our future.

Valorie Aquino
Archaeologist & PhD Candidate
March for Science Co-Chair

When my wife and I wanted to start our family, it was understood that, as a same-sex couple looking to carry our own children, we would need science to assist us. Little did we know how much science would actually be involved. After one intrauterine insemination, seven in vitro fertilization attempts, and my wife contributing her eggs toward our vision for a family, I became pregnant. The road to pregnancy was filled with intense drive, determination, disappointment, heartbreak, lots and LOTS of science, and, ultimately, indescribable joy. This road was also continually guided by love — love for each other, love for the family we were trying to create, and the love from our friends and family who supported us during the journey.

On my 37th week of pregnancy, we attended the Silicon Valley March for Science in Downtown San Jose. We made a sign highlighting how science helped immensely in the making of our baby, with a custom baby bump cut-out to feature my belly front and center. We posted a picture of us to the Facebook group, Pantsuit Nation, and were surprised at how well received it was. We received over 92,000 "likes"; however, what really struck us were the comments posted to the picture.

It is the

MARVEL OF SCIENCE

that has enabled so many of us to start the families we've always wanted.

 Melissa and Serena Cerezo-Poon
Science baby mommies

When I was a senior in high school last spring, I was the lead organizer and MC of the Asheville March for Science. We had a fantastic turn-out at our event with over 1,200 people coming out to support science.

In the days leading up to the march, we still hadn't found a stage. Panic was setting in. Our small team of six called anyone and everyone in our contacts, hoping to find someone with a large platform that we could use.... Anything.

The day before the march, in our 11th hour, a local nonprofit, Asheville Greenworks, saved the day. We hosted the march from the top of a flatbed truck, which was then decorated in green paper from my high school.

This march didn't belong to any one person — it belonged to the community made it happen.

– Luke Shealy
Organizer

MAKE
SPACE
GREAT
AGAIN

T DO WE WANT?

ba sed SCIENCE!

WE WANT IT?

r Review!

MARCH
FOR
SCIENCE

— WASHINGTON, DC —

I grew up
SMACK IN THE MIDDLE
of the Space Shuttle era.

I was the kid with Space Shuttle posters and Space Shuttle LEGO sets, the kid who would sit around and draw rockets all day. Despite this, my love for science and exploration faded by my teen years and didn't revisit me until my late 20's. It was as I watched the last Space Shuttle mission online, STS-135 (July 8, 2011), that I felt a space shuttle-sized hole in my heart. That sense of "Well crap, now what?" led me down a path of learning about the past and future of space exploration.

One day while browsing online, I stumbled upon a used Russian high altitude flight suit on an auction site. I had no intentions of what I'd do with the thing, but next thing I knew, I had a Russian flight suit at my doorstep. After a few months of it taunting me, I eventually started taking funny pictures of myself in the suit, and that's how my project, Everyday Astronaut, was born.

The theme of kid-like curiosity of an astronaut stuck on earth exploring his own planet while waiting for his chance to fly, fueled the inspiration for me to turn Everyday Astronaut into a tool to help people get as excited as I am about the future of space flight. Since purchasing the suit in 2013, Everyday Astronaut has traveled the world, YouTube, and Instagram frontiers as a spaceflight advocate and science communicator.

Science has taught me so much. Most importantly, it's taught me that not only is not knowing the answer ok, but not knowing the answer can be the most exciting thing of all. It means there's exploration ahead, worlds to explore, and questions to be answered.

I find this extremely humbling and exciting. So why space?

I think the 1975 Apollo/Soyuz mission sums up my feelings on why space is so important: In 1975, the United States and our Cold War enemy, the Soviet Union, linked spacecrafts in space for the first time. Our innate human desire to explore came together and bridged the gap between two enemies as we shared handshakes and gifts in orbit.

This is what space represents for everyone who shares this planet together. As we explore space, the Earth becomes home to all of us humans. Humanity grows beyond our imaginary borders, race and wars. It's something we all share together. And as our horizons expand into the great unknown, we do so together as humanity.

 Tim Dodd the Everyday Astronaut
Science Communicator
Spaceflight Evangelist

regardless of **AGE, RACE, RELIGION, GENDER, CASTE, CREED** or even **MUSIC TASTE,** we can all agree that **BLACK HOLES** are the coolest thing **EVER**

Now that's amazing!!

MORRIS, MN, USA

TWIN CITIES, MN, USA

Anyone, including a one-year-old curiously tugging at the tablecloth, can be considered a scientist and should be encouraged in that pursuit.

– *Juan Pablo Ruiz, Bethesda, MD, USA*

PERSISTENCE

IS IN OUR

DNA

Advancing Chicanos/Hispanics
& Native Americans in Science

#SACNASmarches

When I was a Journalism major at the University of Southern California, my advisor told me that I could graduate in three years. Out of fear of entering the real world one year earlier than I expected, I panicked and added a second major: Earth Science. I can now admit that I was just trying to find a set of classes that would be different enough from my normal writing classes that I could catch a break.

First of all, Earth Science was not "a break." I found myself thinking (several times) that I would never pass, science was impossible, and I had made a horrible mistake. Second of all, it was the best decision I have ever made.

I don't find myself using my knowledge about "fission track dating" often, but I did learn how to think like a scientist, which allowed me to understand and communicate any field of science better than many of my journalism peers. I was able to seamlessly communicate with a scientist on their own turf, and quote them about a scientific idea accurately. I made ordinary people understand "magnetic stripping" and I made geologists feel understood.

That is when I realized how valuable it was to know both AP Style and the scientific method.

IT STRUCK ME THAT I COULD DO SOMETHING THAT FEW OTHERS COULD DO. I COULD MAKE "PYROCLASTIC FLOW" UNCOMPLICATED. I COULD MAKE SCIENCE ACCESSIBLE.

And it bothered me that that skill felt like a novelty in both of my fields. It shouldn't be.

Science deserves to be explained well and the public deserves to understand their environment, their food, even their own bodies well.

Three years later (and with an Earth Science degree under my belt) I am the Communications & Marketing Coordinator at the Society for Advancement of Chicanos/Hispanics and Native Americans in Science (SACNAS), a national organization with the vision to diversify the science, technology, engineering, and math fields. Before the march I made a series of posters with slogans from many of our members to print out for the event. Over five months later I still see posters I made in Facebook photos, in internet searches, or even hanging on an office wall.

It's incredible that something I created was part of an international movement that addressed exactly the obstacles I identified in college. More importantly, seeing those posters reminds me that the words, the anthems, and the chants of the March for Science live on.

They are bigger than one event on one day. They are words we should remember every day of our lives, as we raise our children, as we choose our future leaders, and as we fight for science to be a respected, inclusive, and accessible. As a society, that is what we deserve.

Daniela Bernal
SACNAS Staff
Santa Cruz, CA, USA

COLUMBUS OHIO

WHAT I LOVE ABOUT SCIENCE IS ITS FUNDAMENTAL LACK OF BULLSHIT.
Scientific understanding is inherently tentative and incomplete. But the scientific method works! Have an idea, think about it, then go out into the real world and test it.

– Tom Zupancic, PhD
Volunteer

MY NAME IS ERIC VALOR AND I AM THE EMBODIMENT OF THE MARCH FOR SCIENCE.

I taught myself Information Technology by following core scientific principles. I would have something I wanted to do or a problem to fix, would look up possible solutions, devise a plan, and execute that plan. Sometimes it didn't work, but with good planning and careful thought, most often it did. I could then refine that solution for future similar situations.

I was forced to retire in early 2008 due to complications from Amyotrophic Lateral Sclerosis, also known as ALS or Lou Gehrig's Disease. After my retirement, I decided to use the same method to teach myself neurology so I could begin firing my own real bullets in the war against ALS. About five years later I created my own research organization dedicated to finding an effective treatment to stop the progression of the disease. We have a treatment in development right now with what appears to have a real probability of success.

I owe much of my knowledge to the open sharing of information called Open Science. There are many books widely available in multiple formats dealing with particular subjects of Information Technology. As the Internet grew, people were increasingly able to share knowledge and solutions to problems. I, myself, participated in this volunteer work to expand the knowledge of others around the world. So I was uniquely prepared to do what I am doing today — learning about neurology and sharing that knowledge in clear language with my fellow patients who may not be technically or scientifically inclined. They can then make more informed decisions about their treatment options.

It also increases their demand for scientific advance. They know that wishful thinking and miracles are not going to help them — only science can do that. And some have even taken on the challenge of using science to try to find some kind of treatment that can help themselves and others. This resurgence in public participation in science mimics the scientific explosion that shaped the past few centuries.

This Citizen Science has been somewhat forgotten in the past decades as people decided that the mysterious "they" would find solutions to issues and create new technologies. What we forgot is that "they" is actually "us"!

Public participation in science brought us everything from the automobile to the light bulb. The personal computer, perhaps the most influential device in human history, was created in a garage in a suburb!

So we must re-dedicate ourselves to the scientific advancements of human knowledge, ability, and culture. We must not only re-dedicate to this ourselves but evangelize the dedication to others by freely sharing our knowledge with everyone and encouraging each other to do the same.

The image you see is me, as the avatar I created using eye tracking software. With it, I'm able to use my eye movements to type and create a digital persona that moves, talks, and even gives a speech at the March for Science when I cannot.

That's the power of what can happen when science, technology, and human ingenuity meet.

Eric N. Valor
Executive Director, SciOpen Research Group
Former Computer & Network Operations Manager, MBRDNA

Librarianship is a branch of knowledge based on facts and principles. Librarianship has been studied and theorized. Yes, there is an art to practicing librarianship, but at its core, it is a science.

During the 2017 New York State Fair, an informal survey asked people about their library use. Yes, most people reported using a library and some reported using the library frequently. Some were adamant in talking about the importance of the library in their lives. One fifteen-year-old spoke extensively about her use of her library and its importance in her research projects. Libraries continue to be important as a place of facts and information. Libraries remain a place to turn those facts and information into knowledge. Libraries remain a place to seek that which is true in a safe environment.

When people think of libraries, they think of books (paper and digital). Paper would not exist without science. (Go ahead and try to make paper without knowing a formula or method.) Then imagine all of the science used to create ebooks.

In my work as an instructor, what I do would not exist without science. The technologies I use to get work done, including the ways I communicate, would not exist without science. Period.

SCIENCE MATTERS.

If you doubt that, head to your library, where I am sure staff can work with you to locate information to help see the impact of science in your life.

– Jill Hurst–Wahl, MLS
Syracuse, NY, USA

SAN FRANCISCO, CA, USA

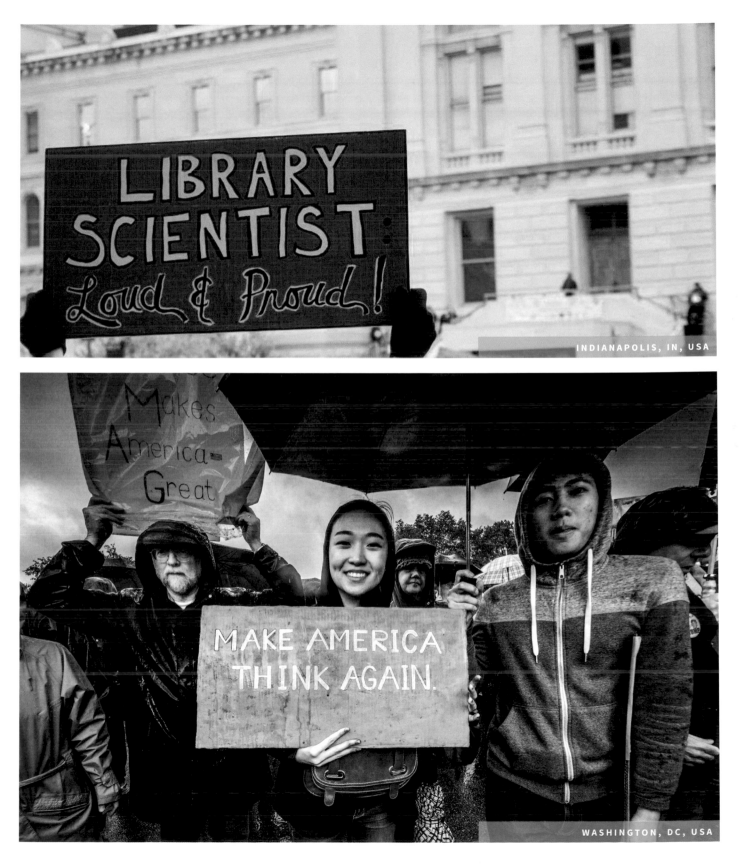

INDIANAPOLIS, IN, USA

WASHINGTON, DC, USA

NEW ORLEANS
LOUISIANA

In order for the march to honor the culture of New Orleans itself, a scientist recommended we have people march in what are called "krewes," so we had different krewes throughout the formation, such as "Krewe of Darwin," which made the marching for such an important cause fun.

Each speaker delivered their impassioned, incisive and eloquent speeches with purpose, compassion, and advocacy for not only science, but for humanity. All of this combined and the honoring of the New Orleans culture, made the march a memorable experience.

– Siobhain McGuinness (Ní Aonghusa)
Organizer

I'M
WITH
HER

SCIENCE → ALCOHOL
ALCOHOL → NEW ORLEANS
NEW ORLEANS = FUN
SCIENCE → FUN

In The Time It Took
You To Read This Sign,
Louisiana Lost 500 Sq Ft
Of Land

TRUST EVIDENCE NOT OPINION

Whenever I have the opportunity to teach others about Native Americans and our desire for a pro-indigenous approach to STEM education,

I begin by simply stating that we have math and science deeply embedded within our cultures and traditional knowledge bases, just as other indigenous people do worldwide.

As native people, we rely upon oral history to impart knowledge. Oral history has not always been considered to be a valid source of data. Because oral history can sometimes be disregarded, most people are not aware that science, technology, engineering, and math (STEM) are not new to us, for they have been part of indigenous cultures for centuries.

Science can be found in ethnobotany through the use of plants for medicinal or artistic use. Biology is present in the agronomy and agricultural techniques of native people, most notably in the practice of planting corn, beans, and squash next to each other. **Technology is found in the manner in which native people used natural waterways to design and utilize irrigation canals for their farms. Engineering is found in the architecture of homes and ceremonial structures — the traditional Navajo hogan or portable teepee. Math is found in the traditional counting systems of indigenous people, through pictures or knots on a counting rope. Anatomy and physiology are taught in the butchering of a deer, sheep, or buffalo. Astronomy and cosmology are also crucial to the way that native people could tell the change of the seasons and even forecast weather.**

I enjoy sharing indigenous epistemologies and I am able to do just that through various indigenous organizations, namely the American Indian Science and Engineering Society (AISES), the National Indian Education Association (NIEA), and the Society of

Advancement of Chicanos/Hispanics and Native Americans in Science (SACNAS), just to name a few. Accordingly, I was honored to be a speaker at the March for Science in Chicago in 2017 to share how diversity drives scientific research. We see the results in the diverse scientific research that is leading the way to address some of the challenges that minority communities face, for example, ensuring that everyone has access to safe drinking water. We also see it in the ways in which we are now democratizing scientific research for underrepresented minority populations, especially for indigenous people globally.

An example of this is bridging the gap between Native Americans and the scientific community in genetic and genomic research. I am Navajo and as an indigenous scientist, I have the privilege of working with other indigenous scientists and researchers to foster a new generation of intellectual leaders (native and non-native) who will define the expanding frontiers of genomic analysis with a specific focus on research with indigenous communities.

This goal is being accomplished in an interdisciplinary learning environment with students and instructors from diverse intellectual backgrounds and a curriculum that empowers Native American students to take leadership roles in efforts to use genomics as a tool for Native American interests. Through this effort, we are not only increasing the number of indigenous scientists and researchers, we are also educating non-native scientists and researchers on how to work with our communities without a savior mentality to engage communities on how to address pressing issues and challenges.

Overall, the major theme in my work as a bioethicist and social justice advocate is to broaden the participation of underrepresented and underserved people and give voice to communities who have not been heard before.

 LeManuel Lee Bitsóí, EdD (Diné)
Bioethicist/Critical Ethnographer
Member, SACNAS

LOS ANGELES, CA, USA

DALLAS, TX, USA

Indigenous science is real science, despite colonization.

Generations of my family have loved the Arctic for over 10,000 years and we fight for it still.

I MARCH BECAUSE MY LIFE, PHYSICAL HEALTH, AND MENTAL HEALTH ARE TIED TO THIS ECOSYSTEM AND ITS MANY INTRICATE FACETS.

I have supported and will support the hundreds of international scientists who are trying to grasp what is happening to our home.

Culture and science are not opposites. I won't technically be marching, but there might be a few of us here in our rural arctic village standing up for science.

– Nasugraq Rainey Hopson
Anaktuvuk Pass, AK, USA

WASHINGTON, DC, USA

CLEARWATER, FL, USA

KAMPALA
UGANDA

In Uganda, our march aimed to celebrate the role science has played, and still plays in addressing critical development challenges such as HIV/ AIDs, hunger, malnutrition, and poverty.

More importantly, we marched to remind our policy makers of the importance of restoring the place of science in major policy formulations and decision making.

We hope to have this march every year as a way of celebrating our pride in science.

– Nassib Mugwanya
Organizer

MY INTEREST IN NEUROSCIENCE CAME OUT OF A CHILDHOOD EXPERIENCE:

LEARNING TO READ.

By age eight, I had fallen significantly behind my classmates. The harder I tried, the less the letters made sense, and the more ashamed I became. My mind would go blank when I was called on. "I can't read," I kept thinking to myself.

Finally, I was sent to a remedial tutor. She noticed whenever my mind went blank, and she would gently coax me out of it. Within a year, I not only caught up, I jumped ahead. A few years later, I started winning citywide writing contests. Still, that rollercoaster experience of falling behind, losing hope, getting saved, and finally succeeding, haunted me. What was different about my brain that I initially failed to catch onto reading? What happened to those like me who never got help?

My scientific curiosity was sparked, and I ended up as an undergraduate at MIT. I majored in Brain and Cognitive Science and studied a region of the brain devoted to learning and memory. In addition, I was interested in using technology to aid learning, so I built educational games at the MIT Media Lab. As college grew to a close, I decided to work in startups that applied scientific research to education and health.

After a few years of doing that in Silicon Valley, I had an idea: what if I could make science — neuroscience in particular — so accessible that everyone could do it? The new, powerful computers in smartphones seemed perfect for the job. I flew back to the East Coast for graduate school at Harvard. There, I built a brain-sensing device for less than $100 using just a smartphone and homemade parts. Also, I designed a citizen neuroscience study using consumer devices anyone could buy, not just neuroscientists in university labs.

At the March for Science, I ran a booth called Neuroportraits. My team wanted as many people as possible to see their brain activity in real-time. We put brain-sensing headsets powered by smartphones on over 100 people that day. We gathered the data in collaboration with Sapien Labs' Human Brain Diversity Project, an international study of brain activity that also included data from people in remote villages in India.

There was one man who came by our Neuroportraits booth that I will never forget. After I positioned the brain-sensing headset, I asked him to solve a short arithmetic problem. Immediately, his facial expression went blank. "I can't do math," he said simply. I felt my stomach drop. It was like seeing the outcome of an alternate, parallel version of myself where math instead of reading had been my struggle. In this parallel reality, however, the necessary help never came.

Although his face looked unemotional, the man's brain activity told a different story. As he attempted to solve the math problem, a waveform typically associated with sleep began to dominate the readout. From the monitor, it appeared that this man's brain sort of "shut down" when confronted with math problems. While I never had brain recordings done as a child, I would not be surprised if my brain activity had, at times, looked very similar to this man's. When I explained that the brain's response to stress can sometimes make it difficult to think clearly, the lights came back on in the man's eyes.

My personal mission is to spread citizen neuroscience tools capable of delivering insights that tell us more about ourselves — and sometimes, that even transform us.

Science not only explains the way we are, it reveals the invisible patterns showing how we could be. I believe that if we understand these patterns, we can write a better future together.

Elizabeth Ricker
Founder, Ricker Labs & Neuroeducate

MR. SCIENCE BEAR

Science influenced me to be a more responsible consumer and mindful of my impact on our environment.

Many years ago, before "climate change" became a political buzzword forecasting the many possible afflictions to our environment, I studied and researched global warming for a class I was taking in college. Back then, the data was not as marred by politics and corporate influence as it is today. I thought surely this pure reporting by academics and noted scientists would prevail to make society change before it was too late!

But now, I see that clear heads may not prevail.

The denial of science is treacherous to all of our futures.

I decided to use my small influence as a graphic designer and marketer to influence others.

Through my Etsy shop, Protest Nation, I design printable posters and postcards as well as t-shirts that allow people to be instant activists. I am trying to do my small part by creating activism art to help others voice their concerns. This design, "Mr. Science Bear", is my personal favorite and I am proud that it has been used across the globe to raise awareness.

 Kristin Black
Graphic Designer, Protest Nation
New York City, New York, USA

We are citizens first, and scientists second. What happens to the world happens to us. This is political, because there are politicians who are making decisions that affect our health, our well-being and our planet.

WHEN WE SEE THEM MAKING DECISIONS THAT ARE NOT BASED ON THE BEST SCIENTIFIC EVIDENCE AVAILABLE, WHEN THEY ARE ACTIVELY, DELIBERATELY IGNORING FACTS, WE WILL STAND UP AND WE WILL PUSH THEM BACK INTO A REALITY-BASED UNIVERSE.

– Melissa Giovanni, PhD
Speaker

Now, more than ever, our scientific voices are vital to the protection of every facet of the environment and science.

In a world gone mad against facts and logic, we scientists must scream against it.

We're ready.
ARE YOU?

– Dayna Collett, PhD
Stillwater, OK, USA

LINCOLN, NE, USA

NEW YORK CITY, NY, USA

RALEIGH, NC, USA

BOSTON, MA, USA

DENVER COLORADO

As an Environmental Health major, I volunteered to write and implement the waste management protocol for the March for Science — Denver. It was a project I worked on all semester. I gathered volunteers from my CSU community. We worked the whole day to ensure the Civic Center Park was kept clean.

It was a rewarding experience, to help make history.

– *Sage Sellers*
Organizer

Without science, I would have never have known how my brain worked. I may never have discovered my ability to strengthen my weaknesses and enhance my strengths. And I may never have realized my power as a voice for the next generation of female scientists.

At 22, I discovered I have a learning disability. I am dyslexic, which means my brain processes information differently than other people's. According to my doctor, it went long undetected because I had found a way to compensate for it unknowingly. Science gave me a way not only to learn about my brain, but to understand what makes it unique.

In many ways, dyslexia has been a gift. It has contributed to my strengths in ways that are hard to describe. Not only did science help me to see that, but my experiences inspired me to take action to encourage young girls to explore STEM fields.

I'm studying to be a Planetary Geologist and know firsthand how challenging it is to be a woman in science. I have faced laughter and criticism for not fitting the stereotypical "look" of a scientist. I know how important it is to be told that you really can do whatever you put your mind to, regardless of what society may tell you. By supporting each other, female scientists can persevere and stand stronger than ever before.

Without science, I would have never have known how my brain worked. I may never have discovered my ability to strengthen my weaknesses and enhance my strengths. And I may never have realized my power as a voice for the next generation of female scientists.

I am graduating next fall from Western Michigan University as the first scientist in my family. I plan to continue encouraging young girls of all abilities to pursue STEM through my science outreach initiative The SpaceGeo Gal. I am proud of who I am and have never felt ashamed to learn that I was different. It gave me the strength to become who I was meant to be and support others as they do the same.

Jamie Marie Ellis
Geophysics Undergraduate Student
SpaceGeo Gal

COLUMBUS, OH, USA

CLEVELAND, OH, USA

In 1980, at the age of 27, my mother (and her then lab assistant, now husband — and my father) made $3,500 a year and slept on the floor in a one bedroom trailer.

They worked 80 hours a week researching the role neuron migration in the embryonic brain plays in certain birth defects. She once told me that she didn't sleep a sound night for three years after buying their first house because they lived paycheck to paycheck. She never knew if we'd have enough money to make it to the next month. They both believed that they would be poor forever and chose their line of work despite that because they loved it and felt that it was deeply important.

Now, more than 30 years later, my mother is the Chair of the Anatomy Department at George Washington University, editor of three major reference texts including Principles of Developmental Genetics, and contributor to countless scientific journals. I know if she had to sacrifice her success for her research, she would do it in a heartbeat. We need to remind this country that scientists do not go into research for profit. In fact, they often go into it despite huge financial setbacks simply because they want to make a difference.

My mother and I march together. I march to defend her life's work. She marches to safeguard the millions of women scientists who will come after her and change the world.

– Rachel Klein
Washington, DC, USA

I'm from a small country town in rural New South Wales, Australia. I majored in criminology and minored in Indigenous studies at the University of New South Wales and I am currently working as a graduate data analyst at the University of Sydney.

I was a part of the March for Science Sydney committee, and as a proud Aboriginal Australian woman of the Dunghutti and Bundjalung nations, I marched for science to promote the awareness of climate change and the impact it has on planet Earth. I Marched for Science to promote the inclusion of Aboriginal Australian knowledge systems and social policy as it is crucial for the future of Australian biodiversity.

Aboriginal Australian cultures are rich in science, having developed over 65,000 years of knowledge of natural, ecological sustainability that was applied to our own lands. The importance of inclusive science is crucial for the future of Australia's sustainability.

I also marched to promote employment opportunities for Aboriginal Australians, as we need more Aboriginal Australian scientists across of the fields of science.

– Tammie Smith
Organizer, Aboriginal Australian Ceremonial Greetings

I had planned on marching in Boston, but my aortic valve stenosis was progressing too quickly.

The surgical team at Hartford Hospital adjusted their schedule to get me in two days before the March for Science.

THEY SAVED MY *LIFE.*

During my recovery, I did a one-man march around the corridors of Bliss 9 East.

Do I love science?
Like never before!

 Mark T. Meyering
Advanced Technical Specialist (Chemistry), 3M inc.

Down With Gravity!

I STILL WANT A JETPACK

Support Science!

TWIN CITIES, MN, USA

The SCIENCE of today is the TECHNOLOGY tomorrow

ST. LOUIS, MO, USA

BOSTON, MA, USA

For me, art and science have always existed together in a very grounded human way.

Both my parents were physicians and they would regularly discuss aspects of how the human body works. They were very passionate about new research as well as their ongoing detective quest to figure out what ailed their patients.

Their inspiration led me to an interest in math and philosophy within the arts, particularly electronic visualization and VR. I sought to understand how we exist in a community and on the planet, through many complex relationships.

After working on a higher dimensional virtual reality grid for one of my theses, I took a job at a planetarium. There, I worked for eight years creating interactive and immersive visualizations with astrophysicists and educators. During that time, I had weekly chats with visitors on what scientific visualization is about, and how we are tailoring the universe to us humans.

Even though everybody may move at different speeds, and some institutions may move backwards at times,

I believe that we are achieving new levels of understanding and connections with each other and with our home planet, enabled by interactive technologies.

– Julieta Aguilera, MFA
Forest Park, IL, USA

THANKS TO ⇨ SCIENCE AND THE HYPERBOLIC COSECANT *Pringles* "AREN'T JUST A REGULAR ELLIPSE"

LOUISVILLE, KY, USA

SCIENTIFIC FACTS ARE OUR ONLY ALTERNATIVE

LOS ANGELES, CA, USA

science is for all Brains

SILICON VALLEY, CA, USA

MAUI, HI, USA

Growing up in Romania In a family of scientists was unusual in the 80s and 90s. For my parents, who are both scientists, doing research with limited resources while raising a child was difficult. I never fully understood how they balanced everything until I had to do it myself. I was determined to have an independent academic career, so I pursued postdoctoral training, and my husband did the same. When our daughter was born, the long hours we both worked at the bench and the trips to the lab we took late at night with her became impossible. I also didn't want to miss seeing her grow up. I left academia, but I was still motivated to help science advance from the outside.

The March for Science took place when I had already been out of academia for six months.

THIS EXPERIENCE REMINDED ME THAT ONE PERSON CAN MAKE A DIFFERENCE IN SHAPING SCIENCE, REGARDLESS OF WHETHER OR NOT THEY ARE IN ACADEMIA.

This thought grew into a more powerful desire to empower scientists to speak up for what works and doesn't work for them and shape their own future in science. While there are multiple issues with the current scientific system, my own experiences made one point clear: we need to work on ways to ensure that scientists who desire an academic career can be successful while also being able to enjoy a nice family life. This way the best and brightest researchers who also happen to be parents can drive science forward from anywhere in the world, and enjoy doing it.

– Adriana Bankston
Louisville, KY, USA

Science is not only my life,

IT'S MY YOGIC PATH.

As a nurse in orthopaedics and an almost twenty-year yoga and anatomy teacher trainer who specializes in chronic pain, I often find myself with one foot in the world of science and one foot in the ancient healing arts. I'm not a scientist, but I am a science educator. I create holistic curriculums and programs for the yoga, medical, and chronic pain communities based on current research.

My work in chronic pain is uniquely challenging, in that there is not a one-size-fits-all answer. As is so often the case with integrative therapy, I'm often the last stop for folks after they've tried everything else. Clients often arrive on my doorstep frustrated, tired, or even angry. It's science that helps me meet their needs.

I've had the great honor of teaching many scientists over the years. I've taught doctors, nurses, teachers, geneticists, and even a few NASA researchers.

One of the things that has always struck me is how much scientists are like ancient yogis. They are infinitely curious, willing to be proven wrong, and constantly seeking answers to the great mysteries of life.

They are often my most thoughtful students, filled with inspiring questions about the things that they don't understand. Their questions challenge me as a teacher and help me to grow on my own yogic path.

Simply put, without the efforts of the scientific research community I wouldn't be able to help the lives that I do.

 Carrie Tyler, RN
Integrative Pain Specialist
Yoga Teacher Trainer

Supporting my hoomans today as they #marchforscience. Mom is a biologist, Dad works in horticulture, but it shouldn't matter what they do because **SCIENCE IS FOR EVERYONE.**

– *Kenton the Dog, IG: @kentonthedog*

Science is FUR Real

OKLAHOMA CITY, OK, USA

ATLANTA
GEORGIA

I believe that the March for Science is important because in a time where a dark, false narrative of "fake news" and "alternative facts" is being perpetuated, this movement shines a light with facts, knowledge, and truth. It inspires and welcomes all to become informed, educated citizens about the reality of our world, and to take meaningful actions towards making our planet a healthy and safe place for all.

Photographing the science march was a very inspiring experience. I captured many people across all different walks of life, all unified by one goal — the promotion of the importance of science, facts, and taking care of our beautiful planet. I feel fortunate to have photographed so many passionate, bright, and magnetic people coming out and fighting for what they believe in.

And I'll be entirely honest, photographing the young kids holding up signs about wanting to become future scientists, engineers, and doctors, absolutely made me teary. I hope they never lose that fire, and never stop believing that they can become anything and change this world for the better.

– Catherine Marszalik
Photographer, Athens, GA, USA

My love for physics
STARTED WITH WATER BALLOONS.

My love for physics started with water balloons. In high school, my physics teacher showed me how to calculate the trajectory of a water balloon fired from a sling shot. I was hooked.

I went on to study particle physics, eventually earning my PhD. Yet physics couldn't provide answers to all of the problems that were important to me. In between my early studies and lab work, I took to the streets to organize my community in our fight for marriage equality. When I moved on to graduate school, balancing both efforts became too much. I put my activism aside to focus on my research. **That didn't last long.**

Two years later, something was missing. I began questioning whether a career in academia was right for me. I switched gears again, this time shelving my science completely to direct campaign offices during the 2008 presidential election. Despite the amazing friends I made and exhilarating experiences I had, I still felt a longing to be part of the scientific community. Confused and unsure, I decided to return to finish my PhD and sneak in community organizing where I could.

It wasn't until the end of graduate school that my path finally became clear. I received an email from my advisor about a science policy fellowship. Was "Science Policy" an actual thing? Had my meandering path uniquely prepared me for a career I didn't even know existed? Maybe I didn't have to choose between science and policy after all.

Three days after my final physics presentation, I headed to Washington, DC, to explore this possibility. I was fortunate to receive an internship at the Union of Concerned Scientists and a fellowship at the National Academy of Sciences. Finally, I received the fellowship that I had been dreaming about since my advisor sent me the advertisement: working in Congress as a AAAS Fellow.

Early on in my time in DC I met with a senior congressional staffer. They told me that working on the Hill is a lot like working in science: you form a hypothesis and then go out and find data to support it.

Only, that's not how science works.

In science, we don't go looking for data to confirm what we think will be true. Used correctly, the goal of science is actually to defend against that impulse. We collect data to test a hypothesis, allowing for the possibility that we were incorrect. If that happens, we often need to change the way we think about a problem.

The staffer's misstatement of the scientific method led me to a critical realization: science can and should play a role in public policy. Scientists train their whole lives to do objective analysis that allows for changing perspectives, which is sorely needed in our current political climate. But there's work to be done on both sides.

Scientists need to realize that science is one of many important inputs for public policy. We also need to do more than just create knowledge; we need to make sure that knowledge makes it into the hands of the people who can do something with it.

And policymakers need to do their part when scientists show up. They need to appreciate the scientific method and set responsible standards for incorporating evidence into policy — even when the data don't fit their hypothesis.

Dan J. Pomeroy, PhD
Managing Director & Senior Policy Advisor, MIT International Policy Lab
Former American Geophysical Union Congressional Science Fellow

WE LIVE IN A SOCIETY EXQUISITELY DEPENDENT ON SCIENCE AND TECHNOLOGY, IN WHICH HARDLY ANYONE KNOWS ANYTHING ABOUT SCIENCE AND TECHNOLOGY.

~CARL SAGAN

LITTLE ROCK, AR, USA

TEACH SCIENCE

$E=mc$

MIAMI, FL, USA

POWER to the TEACHER

math science

COLUMBUS, OH, USA

Vacations with my parents greatly influenced my love for learning about science. We've traveled to Hawaii several times. Visiting Volcanoes National Park and up-close experiences with sea turtles are chances to see science in action!

Science was always my favorite subject growing up, in part because I had such outstanding science teachers in elementary school and high school.

Now as an elementary school science teacher myself, it's my goal to help students see themselves as scientists and discover how and why the natural world functions the way it does.

– Meggan Romei
Chicago, IL, USA

WASHINGTON, DC, USA

SCIENCE WAS MY FIRST LOVE,

BUT THROUGHOUT MY LIFE, I'VE SEEN ITS RELEVANCE QUESTIONED.

Growing up on the southside of Chicago, it was very odd to be intrigued by all things STEM. I was ostracized from my peers at an early age due to my eccentric interests. My desire to learn about microscopes and atmospheres was weird and wrong. I felt insecure in my fascination about science, and eventually began to suppress it.

As I've matured and reintroduced myself into STEM, I have become certain that science advocacy is a dire necessity. I now realize that people who look like me suffer unnecessarily due to lack of science literacy. I've taken this problem personally for as long as I've been aware of its existence.

I never truly felt supported in my stance until I attended the March for Science.

Marching alongside Bill Nye and other major influencers in the science community showed me that this lifelong pursuit does not have to be solitary. Seeing the global response to this mission's urgency inspired me to quit my job and start a nonprofit, Fascinate, for science advocacy.

Fascinate's mission is to excite underrepresented youth about STEM. We have begun raising funds for the #MagicCoolBus because we believe that a cure for cancer, or solution to world hunger, may be trapped in the mind of a child without the resources to reach his or her potential.

 Justin Shaifer
Executive Director, Fascinate
Science Communicator

WASHINGTON, DC, USA

WASHINGTON, DC, USA

SILICON VALLEY, CA, USA

LANSING, MI, USA

CHICAGO, IL, USA

When people ask me where I'm from, I laugh. I was born in Canada, raised in Italy, attended college in England, and am currently a PhD student in the US. I am a citizen of the world, but I am a scientist for the world. Science isn't perfect because people aren't perfect; yet it is the reason for the medicines, wisdom, adventures, and technology that make the US — and the world — great.

It is our responsibility as citizens of humanity to champion progress towards the greater good, to support inquiry, to celebrate discovery, and to commit ourselves to building an equitable global society that allows everyone to thrive.

– Gabriella Hirsch
Chicago, IL, USA

HIGH · FIVE · FIVE

FOR

Science

SCIENCE IS SUPER

As a designer, my job is to take complex, nuanced ideas and give them a voice that is quick, understandable, and shares a narrative with the viewer.

That's also my job as a comic creator and artist; finding the strongest way to get across my message to the widest audience in a way that involves them. I have always been fascinated with the ideas and design of propaganda, and as I studied it, I wanted to see the positive power of propaganda techniques convey strong, uplifting, and powerful messages instead of divisive, phobic, and negative ones.

As a storyteller, I make stories that are populated by smart, driven characters who are diverse and passionate about what they stand for.

THE MARCH FOR SCIENCE'S MISSION ALIGNED WITH MY DESIRE TO SEE DIVERSITY AND RANGE IN A FIELD THAT CAN BE MADE STRONGER BY INCLUSIVE AND BALANCED PRACTICES.

As a visual communicator, I refuse to let the powerful tools I have at my disposal not be used to give a loud, confident voice to ideas that deserve a wider broadcast. We live in a world with a lot of noise; hopefully, my work boosts a positive signal.

 Paul Sizer
Owner, Design Director, Sizer Design + Illustration
Kalamazoo, MI, USA

The March for Science in Bozeman, Montana had nearly 1,100 participants. That's approximately 2.5% of the population of Bozeman! Although other communities had greater numbers of participants, I think our march had one of the higher per capita rates of participation. So, that tells me even people in the hinterlands are standing up for science.

– Robert K. D. Peterson, PhD
Speaker & Organizer

BOZEMAN
MONTANA

kids in tech

presenta

Como hacerse un "blogger"

BLOG

¡Enseñe escribir artículos eficaces y diseñe su propio "blog"!

Dónde: SII Program at Coalition for a Better Arce 517 Moody Street, Lowell, MA

Fechas: julio 2, julio 18, julio 19, julio 26, julio 27, agosto 1, agosto 2,

I SET UP MY FIRST COMPUTER WHEN I WAS VERY YOUNG.

My parents were immigrants and though they worked hard, we were struggling at the time. My dad was working on his PhD and using computers to code and do calculations for his research. It dawned on him that these machines were going to be a big part of his children's futures, so he saved his money to buy one for our family.

My dad has always been a very practical man and he took every opportunity to steer us towards successful careers. He decided that we weren't just going to get a computer; we were going to set one up ourselves. We talked through the steps, learning together which piece went where and why. And when it was finally done, my sisters and I would fight over who got to play on it. It wasn't until many years later that I looked back and realized what a pivotal moment that was in my life.

I didn't just know how to use a computer — I knew how it worked. And that was the catalyst for a lifelong empowerment and comfort with technology that not all kids have access to.

Today, I'm an educator by formal training, but I never lost my love of technology. I've always been around scientists of some sort and seen how technical skills can help advance so many careers. I also know now that mentors and project-based learning can go a long way for a kid. I am thankful for that experience.

As I got older, I realized that not everyone had a mentor like my dad or an opportunity to feel empowered to tackle hard questions or set up something themselves. Science or technology plays a role in just about everything we do. Science is just about discovering and organizing knowledge to get things done in the world, and technology is one of

those things it can do. As these forces shape the world we all live in, it's increasingly important that people of all backgrounds understand the role that science and technology play in our day to day. We need to be working towards a future that is more inclusive in all senses of the word, especially for women and people of color in the STEAM (Science, Technology, Engineering, Art, and Mathematics) fields. And the earlier we start showing kids what they can do, the better.

So I started Kids in Tech, a nonprofit focused on creating pathways and an early entry point to the STEAM fields. We serve under resourced communities as they build skills in computer literacy and computer science. In many ways, it's modeled after my own experience. We give kids the tools they need. We give them projects that teach them the importance of persistence, but also instill a sense of ownership over technology. We surround them with mentors that believe in them and provide a community to help them grow.

Access to technology is about equity. The empowerment that comes from those early lessons is what leads people to become scientists or engineers, or to appreciate the role of these things in their lives. It's about opening up the world and letting our kids become the creators so that they can thrive in the 21st century.

I'm very grateful to my dad for showing us at such a young age what we could do; all because he thought those skills would come in handy later on. He was right — and now I plan to replicate that moment for as many young kids as I can.

Olu Ibrahim, MEd
CEO and Founder, Kids in Tech, Inc.

NEW YORK CITY, NY, USA

LONDON, UK

Science saves lives. Science teaches us to live.
Science is probability. Science is a new horizon of prospect.
For any problem, science has solutions.
In creation, not destruction, in light not in darkness.

– March for Science Bangladesh
#ScienceAlly

It can be easy sometimes to see and hear people advocate for a cause but not see how they personally relate to it. And that personal connection can make all the difference in the world.

I am the CEO of Oakland-based nonprofit Dream Corps and Executive Director of Green for All, Dream Corps' internationally renowned environmental advocacy initiative. I don't mind saying that I am an environmental policy expert and strategist, that I work to build an equitable green economy, and that I have developed numerous energy, environmental, transportation, and economic policies and programs at state, federal, and local levels. I also know that to properly advance sound science in a practical way, we have to see how communities are affected and hear from community members what they need.

I was raised in Oakland, a community that has suffered greatly in poverty, violent crime, and a variety of harmful environmental factors. I am the mother of twin boys. And as I travel the nation, visiting other communities that are hurting from the negative impacts of climate change and suffering because their living conditions have been weakened by inadequate or non-existent environmental protection policies, I see and hear things that could harm my sons one day.

Green for All recently launched Moms Mobilize, a national campaign to inspire and "activate" moms to join forces and raise their voices and influence their communities for the environment, for the sake of their children.

In my travels, I met eleven-year-old Jaden in New York. Jaden, who was born three months prematurely and suffers from a rare form of diabetes that makes him constantly thirsty, is the author of "Kid Brooklyn," a comic book about environmental activist heroes who combat a corporation that tries to dump harmful chemicals in public water sources. You can view a video of Jaden at the Green for All website.

I have also met moms in New Orleans, in Oakland, where I was raised, and in Flint, Michigan. And no matter where you live, what we share in common is the desire for our children to have clean water to drink and bathe in.

CLEAN WATER IS A RESOURCE THAT SHOULD BE A GIVEN. IT ISN'T.

And it is so lacking in some parts of our country that there is a direct correlation to poorer health for children. Poorer physical health often leads to poorer academic performance, which can lead to exponentially fewer opportunities. This sort of domino effect harms low income communities disproportionately, particularly communities of color.

But this fight isn't just about clean water. As recently as last year, the American Lung Association ranked the ozone quality in Flint as among the worst in the nation, and unaddressed infrastructure issues have contributed to a dipping population and climbing poverty rate.

My work can't be done without passion and compassion. It can't be done without personalizing the fights we all face. It can't be done without recognizing the connection between a healthy environment and improved economic opportunities for historically depressed communities.

And it can't be done without sound science to educate the public and hold governments accountable for the health and clarity of our air, water, and land.

 Vien Truong
CEO, The Dream Corps
Director, Green For All

MIAMI, FL, USA

PORLAND, OR, USA

I grew up in a small town in Montana where there were only 300 people. My graduating class was 26 people. We didn't have a lot of resources, but my teachers did everything they could with what we had. Growing up in a small town I also never met an out gay person until my junior year of high school when I got to go to Washington, DC, for the Close Up program. I'm now an out gay man getting my PhD in Chemistry. I march to show that even a small-town gay boy from Montana can become one badass inked astrochemist.

– Alec Kroll, Boulder, CO, USA

THIS IS WHAT A
CIENTIST LOOKS LI

Mrs. Michielsen

Mary Lois Michielsen, Mother of 7
M.S. Chemistry, 1969

THIS
is WHAT a
SCIENTIST
Looks Like

HOMER, AK, USA

AUSTIN, TX, USA

I march to show every kid, no matter where they are from, who they are, or who they love, that **SCIENCE IS A DIVERSE PLACE THAT BENEFITS FROM THE PRESENCE OF ALL OF US.**

All right, but apart from Safe Drinking Water, Clean Air, Plentiful Food, Longer Lives Drastically Lower Rates of Infant Mortality, Cancer Treatments, The Eradication of Polio, The Internet, Cell Phones, Cars, GPS, Space Exploration T-Shirts That Don't Shrink in the Wash, What have the scientists and engineers ever done for us?

Science doesn't just create jobs for the people who do it, it creates entirely new job sectors. The way we travel, the way we produce food, the tools we use to teach, our entertainment, our understanding of ourselves and each other: they're all shaped by science. I think people forget what science really is. It's our working model for what is possible.
MORE SCIENCE, MORE POSSIBILITIES.

– Maya Bialik
Cambridge, MA, USA

BOSTON, MA, USA

SAVANNAH, GA, USA

LOUISVILLE, KY, USA

STEM**INIST**

I started Beyond Curie, a series of illustrations and stories that celebrates the rich history of women kicking ass in science, technology, engineering, and mathematics,

SO THAT YOUNG WOMEN EVERYWHERE COULD STOP WONDERING IF THEY HAD THE POTENTIAL TO MAKE AN IMPACT IN SCIENCE AND INSTEAD ASK THEMSELVES, **"WHY SHOULD WE STOP NOW?"**

It is about discovering our heroes.

Because when we connect with greatness we come to see that it is not some distant unreachable place, but a long body of work forged through perseverance, love, and courage.

Amanda Phingboddhipokkiya
Creative Director and Design Strategist
Brooklyn, NY, USA

FOUR YEARS AGO I WAS LOOKING FOR A CAREER CHANGE.

I was ending a path in the social services industry to find a new start in a completely different field. My friend suggested I try wildland firefighting given my love for the outdoors.

As soon as I started with my first crew, I fell in love with the work. There is something immensely satisfying about long days of physically demanding work side by side with people you consider more like brothers than coworkers.

We get to see the savage beauty in a force of nature few people will ever experience. It's also a field that values continued learning. No one has enough experience to no longer be a student of fire.

Wildland firefighting involves long hours of intense physical labor. However, many of the safety precautions, planning, and tactics we use are based on the science around fire behavior. We designate one of our crew members to sling weather every hour, sometimes every half hour, to monitor how the weather is affecting fire behavior and to help predict what the fire will do. Science and technology let us achieve this through tools to measure pinpoint accurate weather readings for temperature, relative humidity, wind direction and speed, fine dead fuel moisture, and probability of ignition for the specific area in which we are working.

Every geographic area has its own threshold for temperature, humidity, and wind that allows for a significant increase in fire intensity. Once we observe that one of these predetermined thresholds has been met, we stop, observe, reevaluate, and determine the best way to safely reengage or pull back and adjust our tactics. These weather observations are a critical component of what keeps us safe in our efforts to contain a wildfire.

And it's science that makes it possible.

 Jeff Ellis
Firefighter

To most people I am credited as being a scientist because I have a PhD in neuroscience. I have been formally studying various aspects of brain function and behavior since I was a junior in high school.

It isn't my degree, published manuscripts, or expert knowledge alone, however, that makes me a scientist. Rather, it's my curiosity and desire to understand the natural world that allows me such a designation.

For as long as I can remember, I have experienced daily life as one science lesson after another. As a little girl, something as simple and practical as making a pitcher of Kool-Aid with my dad was a chemistry demonstration, because I would proudly recite the property of crystals and chemical bonds to my mother as I stirred the sugary, liquid mixture. Learning to drive as a teenager was a crash course in physics, as my dad emphasized the laws of motion for me to avoid crashing his car.

To date, I don't think any of my formal science education was nearly as relevant or entertaining as it was with my dad as an instructor. Because with my dad, science was more than just a concept in a book, but a manner of explanation that could be directly applied to any- and everything. There's something really thrilling and satisfying about understanding and explaining the world around me, whether in the lab or in everyday life. Because science is something that we do, not just something that is.

I think of the word "science" as more of a verb than a noun. And while I officially science from 9 to 5 for my profession, I also science when I'm at home, in my periodic table of elements leggings, when Googling NASA's latest missions into space, or while explaining on Twitter why our emotions aren't controlled by our physical hearts.

Sciencing is an attitude and a way of thinking that anyone can take part in. It is my hope that as many people as possible feel empowered to science, to explore, to comprehend the world around us. Because the more we know, the more we share, and the better we will science for the generations to come.

 Marguerite Matthews, PhD
Science Policy Fellow

W/O SCIENCE we'd be Harry without a Hermione.

KANSAS CITY, MO, USA

without SCIENCE it's just FICTION

MADISON, WI, USA

Only Decepticons use Alternative Facts

TALLAHASSEE, FL, USA

CENSORING SCIENCE IS WHY KRYPTON EXPLODED

WASHINGTON, DC, USA

WE SHOULD BE BUILDING STARFLEET NOT WALLS!

SAN FRANCISCO, CA, USA

THE CREDIBLE HULK always cites his → sources ←

SAN FRANCISCO, CA, USA

Join The Dark Side

Lack of Science Disturbing

NEW YORK CITY, NY, USA

Kid Power

TWIN CITIES, MN, USA

I was so happy that we got thousands of passionate people out to support a worthy cause: science. Here in Texas, thousands decided that supporting hypothesis-based testing and decision making was something they couldn't idly stand by and watch be dismantled by the federal government. I'm proud of our March — we did it our way with people of all ages, ethnicities, and belief systems turning out to march for a common goal of a commonsense future.

We continue to be vigilant in opposing policies that violate our beliefs all the while continuing to keep up the momentum on a local level that will promote policies that have evidence backing their efficacy. We look forward to the movement continuing to back politicians (federal and local) that support our worthy cause. It's only through science and evidence-based problem solving that well be able to tackle the largest problems of our time.

– Daniel Barros
Administrator

DALLAS
TEXAS

I remember like it was yesterday the moment I decided what I wanted to be when I grew up. Not an astronaut, not a firefighter, and no, not Xena (okay, I'll admit that a small part of me desired to be this badass Warrior Princess).

I WANTED TO BE AN "ANIMAL RESCUER," LIKE JEFF CORWIN.

While I didn't know it at the time, I actually wanted to be a scientist: a wildlife conservationist, to be exact.

Fast forward ten years later and my love for nature and animals was still going strong. After graduating from college with a BA degree in Community Studies (social justice and community organizing), I volunteered in Monteverde, Costa Rica, at Cloud Forest School, a bilingual, environmental education school. There, I worked to educate the children about the natural world and the important role it plays in our lives.

My brief time in Costa Rica reconnected me to science, to nature, and to my place in this world. It reaffirmed my life's purpose to inspire and co-create positive global change as a steward of the earth. I'll never forget my magical encounter with a spotted paca as I sat quietly on a log deep in the cloud forest. I froze, in complete awe and wonder as I stared at this creature that looked like a giant guinea pig, so close to me it seemed as if I could reach out and touch it.

I returned home to California with a newfound appreciation for nature. My perspective had shifted and my passion reignited, leading me to an internship with Island Conservation, an international organization that prevents extinctions by removing invasive species from islands. The invaluable experience working there and with other nonprofits led me to my job as Social Media and Communications Coordinator at SACNAS, the largest multicultural and multidisciplinary STEM diversity organization in the country. Never in my wildest dreams did I anticipate leading SACNAS's

participation in the March for Science by mobilizing our membership and marching in solidarity with our diverse scientists of color.

I marched for our ancestors, for biodiversity, and for future generations. Most importantly, I marched as a vegan activist. I belive that choosing a plant-based diet is the single most powerful action I can take to reverse climate change and co-create a more peaceful and sustainable world. It is the rent I pay for living on this planet. It is at the core of my identity as an environmental activist.

When I was a little girl, my favorite animal was a polar bear. I'm still obsessed with these majestic white bears that live on the ice. I can't imagine my children growing up in a world where polar bears and other magnificent creatures do not exist. Science has compelled me to devote my life to healing Mother Earth.

This is why I plan to be vegan for the rest of my life and also raise my children vegan. I want to teach them that a plant-based diet allows us to coexist with the other species we share this earth with.

We all have to find the things that we can do to address the issues that science reveals. I plan to continue my environmental activism as a vegan chef, cooking class instructor, and writer. I'm blessed to be a part of this growing plant-based movement and am excited to see where this journey — and this movement — takes me next.

Chiara Cabiglio
Vegan Chef and Conservation Activist, Conscious Eats for Life
Former SACNAS Social Media and Communications Coordinator

I march with my dad. In 1961, he was lucky enough to be the only person in his village in India to gain admission to an engineering college. He barely had enough money for the train ride to the college town.

From there, he gained admission to a graduate science program at the University of Cincinnati. He had to get a special grant from the government of India to afford the plane ticket to the US.

He was given a stipend of $115 a month — barely enough for food — but a chance to build a science career. And he did — one that took us from Buffalo, NY, to Tulsa, OK, to Roanoke, VA, to Butler, PA.

He ended up starting a business — one that lasted 30 years and employed hundreds in rust belt towns in Pennsylvania.

His story is the story of millions of immigrants to this great country.

He's now 80 — dealing with health issues that will soon take away his ability to walk.

So I march with my dad — the greatest scientist I've ever known — to honor all that science made available for me and to ensure that opportunity remains for immigrants in the future.

- Kishore Hari
San Francisco, CA, USA

SAN FRANCISCO, CA, USA

APPLETON, WI, USA

SEATTLE, WA, USA

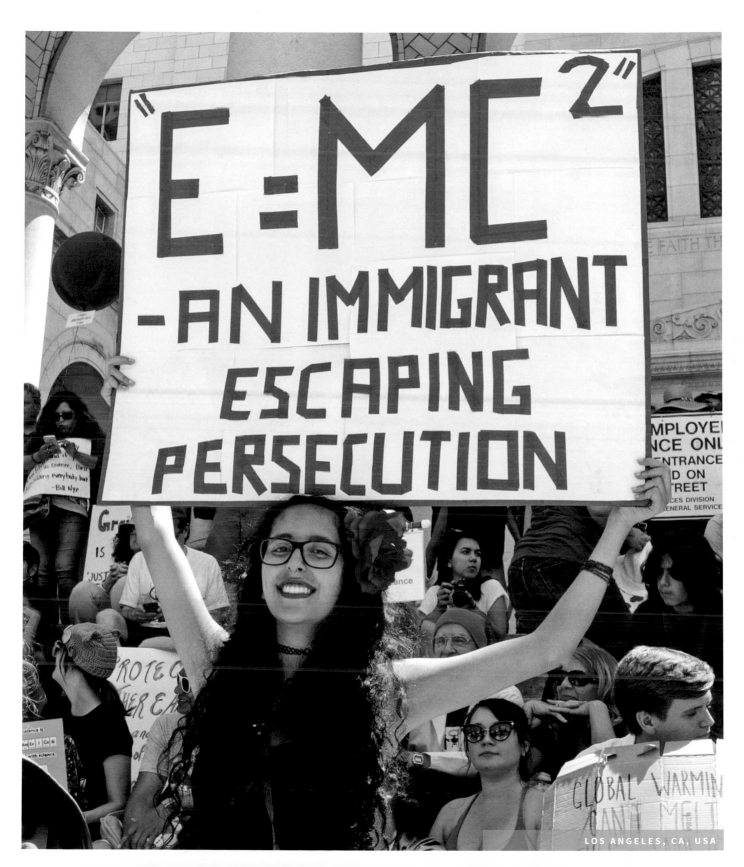

"E = MC²"
– AN IMMIGRANT ESCAPING PERSECUTION

MINN

ESOTA

I MARCH TO
Protect the Basics
WATER, AIR AND
Food!

I am marching
for Mental
Health Awareness

I learned science is always in progress, and it opened up a whole new world of possibilities.

I would never have thought that I would end up here, working in a microbiology lab on a few very cool projects.

In high school, I wasn't exposed to the many aspects of science out there. I thought if you wanted to pursue the sciences, you had a few options. You could become an engineer. Or a software engineer. If you wanted to study life, you could major in biology and go into some health profession, like a doctor, a pharmacist, or a vet.

In college, I quickly learned that this was not true. I started working in a microbiology lab in the summer of my freshman year — taking advice from friends who said that it could build up my resume — and my career path was changed forever. I worked under Dr. Mary Firestone, an esteemed researcher in soil microbial ecology, whose lab was filled with pleasant, encouraging people who gave me many opportunities to grow.

I learned that there were people out there making the next wave of discoveries that would end up in textbooks some day. That others were developing new technologies with biology (no medical school degree required)!

I learned science is always in progress, and it opened up a whole new world of possibilities.

It was here that I gained a fond appreciation for the little microbes who do so much in the world, yet can't be seen by the naked eye. I was amazed to learn that people can take small samples of soil and use microscopes to figure out the tiny species that live there, and to learn that microbes actually help plants grow. Furthermore, I was intrigued by how different day-to-day tasks could be as scientists — one minute I would be working with DNA, and the next I would be transplanting germinating seeds into soil to grow into healthy plants.

I am still trying to figure out what I want to do in the long run, but I know I want to continue pursuing science.

In the microbial realm, there is still so much we have to discover. I would like to be able to either discover new things about the microbes of this world or to figure out how to use microbes to help humans in practical ways. I would like to continue inspiring people of all ages with educational outreach and mentoring so that they, too, can be as motivated as I was to pursue science.

I have high hopes for the generation to come, hopes for a generation that will help us find out more about our world. And make really cool things!

Jessica Trinh
Research Assistant

Wielding of knowledge is what science is all about.

Understanding our problems and issues well enough to solve them is the goal, and engaging in wisdom partnerships will strengthen us all and get us there together.

E ALU PŪ. IMUA!

[Let's combine efforts. Forward!]

– *Sam ʻOhu Gon III*
Maui, HI, USA

SAVANNAH, GA, USA

LOS ANGELES, CA, USA

PHOENIX, AZ, USA

SCIENCE
NOT SILENCE

MARCH FOR SCIENCE 4.22.17

I BELIEVE IN SCIENCE

A rocket blasting through space is such a recognizable, un-ignore-able example of what science can accomplish.

I think when you see something like that, and marvel at it, it cracks open some curiosity, and even some hope. It's this soaring, roaring reminder that, even amidst all our arguing, we can keep exploring, and keep asking questions, and keep finding answers.

I'm no scientist. But I believe in science.

I like knowing that problems are being studied, by people who are careful and smart and working together. I like trusting that a lot of those problems will eventually get solved. I like trusting that, somehow, dumb decisions aside, we still have a chance to save this planet.

And I do feel like science is under attack.

When you see the potentially irreversible destruction of our ecosystem portrayed as some kind of "hoax," it's time to stand up. I hate hearing people say that if scientists don't all agree on something, none of them knows anything. It scares me when people decide that agreeing with their friends matters more than aligning with the facts.

As an artist, I feel like I have a chance to show my support in a visual, shareable way.

A pin is perfect for that. My hope, along with my partners in Pincause, was to give all of us who value science a powerful image we can hold in front of us — and stand behind.

Penelope Dullaghan
Artist
Indianapolis, IN, USA

WASHINGTON, DC

The night before the march, I was seized with fear that no one would show up — that the crowd would consist of me, my fellow organizers, and my parents. On April 22, I watched the crowd swell to more than 100,00 people in the pouring rain and realized I should never have doubted the power of science to unite us.

The Marches for Science sent a powerful message that people around the world value science, but they were only the beginning.

– Caroline Weinberg, MD
March for Science Co-Chair

Cancer doesn't care about race, religion, sexuality, nationality, or socioeconomic background. **Why should the treatment of cancer depend on such factors?**

I am a physician and a cancer scientist and I was inspired to March for Science by the hundreds of thousands of people who mobilized to support and protect what my colleagues and I do.

I study the effects of cancer on the immune system. Recent astonishing advances in science have rendered some of the most lethal cancers treatable by reprogramming our immune systems to target cancer cells inside us, or by unleashing the breaks on the immune system that cancers exploit to hide within us. However the costs associated with these therapies render them limited to those lucky enough to have access to an economically developed healthcare system.

But cancer doesn't discriminate; cancer doesn't care about race, religion, sexuality, nationality, or socioeconomic background. Why should the treatment of cancer depend on such factors? We have already come so far in the way we treat cancer. It is so immensely important that we continue collaborating with researchers from all parts of the world to find new and improved and economically viable treatments for these devastating diseases.

I marched because I hope to help answer some of these challenges within my lifetime, and if I do so it will be because I was standing on the shoulders of giants, and cheered on by the spirit of thousands of passionate protesters.

Athalia Rachel Pyzer, MD, PhD
Physician Scientist

SAN FRANCISCO, CA, USA

WASHINGTON, DC, USA

WASHINGTON, DC, USA

SAN FRANCISCO, CA, USA

HOUSTON, TX, USA

OMAHA, NE, USA

RALEIGH, NC, USA

In Mexico, relative investment in science and the arts is very low in comparison to other countries, like the US. This situation is not due to lack of money. Several of richest men in the world live in Mexico while a third of the population lives in poverty, reflecting the gross economic inequalities. Every year more people are graduating in sciences but the workplaces and jobs are scarce, because of the low interest of foreign companies and Mexican entrepreneurs in investing in original research in Mexico. The sum of these conditions results in a grim picture for people like me, who want to live in a world that supports science and research.

The Mexican government not only holds corruption as a banner, but stands out for its ignorance. It has funded the national institute of homeopathy, approved the use of seers and psychics to solve murder cases, and supported pseudo-scientific projects like research on "human photosynthesis." Against this backdrop, a decrease of 27% of the budget assigned to science was announced at the end of 2016. These are just some of the reasons why I stepped out of the lab and took to the streets — to participate actively in the decision-making of my country. It is fundamental to understand that political participation is not restricted to voting. For me, personal and collective reasons form an amalgam.

I MARCH TO LOOK FOR BETTER SOCIAL CONDITIONS AND I THINK THESE WILL ONLY BE ACHIEVED THROUGH SCIENCE.

– Adhemar J. Liquitaya Montiel
Organizer

MEXICO CITY
MEXICO

Science Will Help my Mommy Walk With ME

In 2010, I was playfully pushed into a pool at my bachelorette party and sustained a C6 spinal cord injury. I am unable to use my fingers, my triceps are very weak and I am paralyzed from the chest down. I dream of walking with her at the beach through the sand and teaching her how to swim. I'd love to get on the playground with her and run around.

Unfortunately, every year, 12,000 people will experience this injury. Some will recover but many will be left paralyzed unable to do the things they once loved.

PEOPLE ALWAYS TELL ME THAT ONE DAY I WILL WALK AGAIN BECAUSE I'M SO STRONG AND INSPIRING. I ALWAYS TELL THEM THAT HARD WORK ISN'T GOING TO REPAIR A SPINAL CORD INJURY; **ONLY SCIENCE WILL DO THAT.**

Rachelle Chapman
Mom and wife

**GOOD SCIENCE REQUIRES FREE INQUIRY AND GOOD INFORMATION, NOT DOCTRINE.
BETTER INFORMATION MAY OVERTURN EVERYTHING WE KNOW.
REMEMBER GALILEO.**

– Bruce Degen
Illustrator & Co-creator of Ms. Frizzle

LOUISVILLE, KY, USA

NASHVILLE, TN, USA

If you keep an open mind, you never know who might walk in!

think CRITICALLY, kids!*

*and adults too!

WASHINGTON, DC, USA

RALEIGH, NC, USA

OAKLAND & SILICON VALLEY, CA, USA

A's
WARRIORS
#intersectional
#futurist
Decolonize
Science
Center
Women of Color
#MarginSci #LGBTQIA

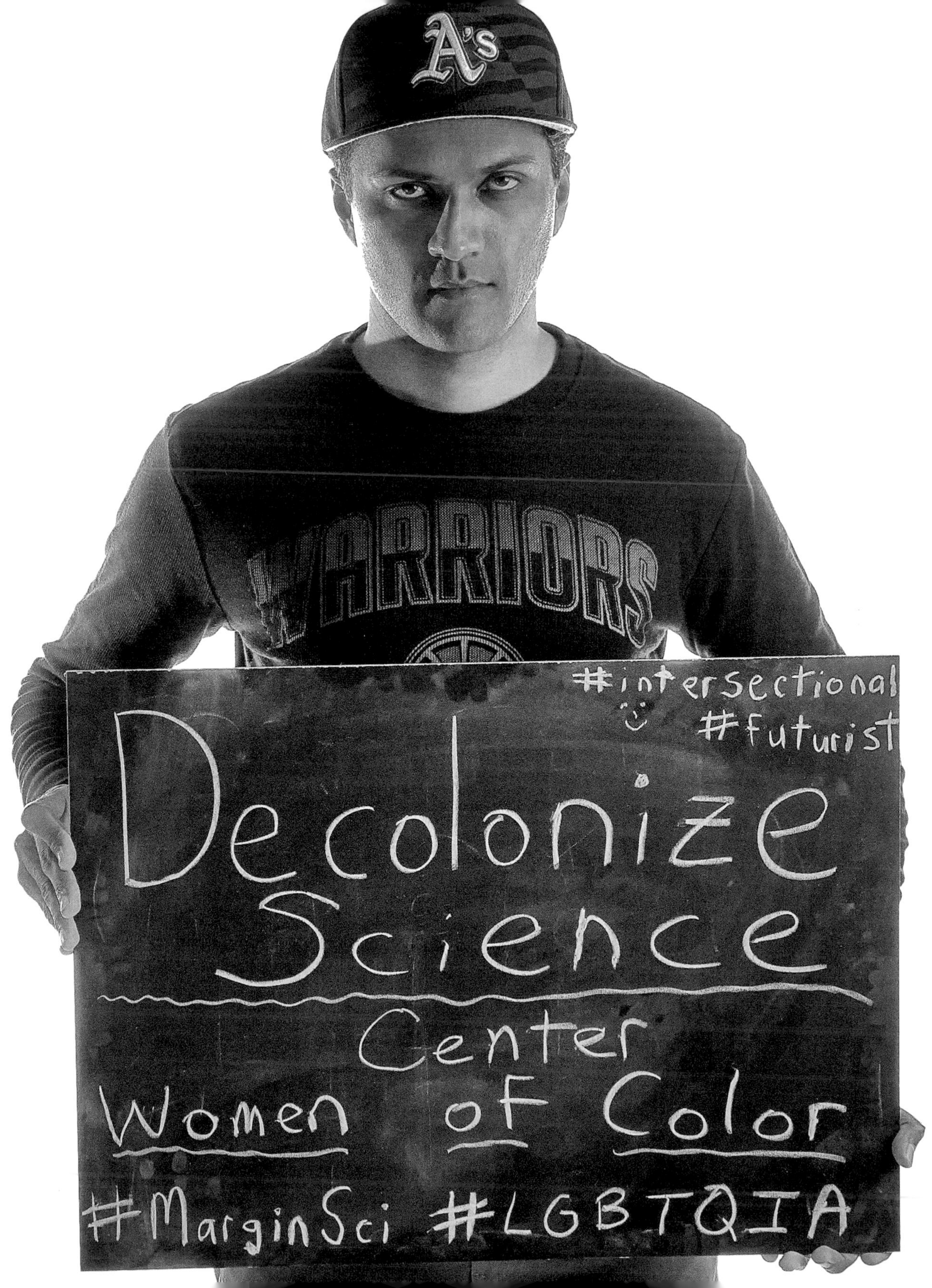

THIS IS ABOUT **CONNECTING** WITH **EVERYONE** WHO FEELS THAT **SCIENCE IS** SHUTTING THEM OUT.

Ravi Valleti
Futurist Author, "Rocket Scientist"
Lead Actor, "Devised"

One of the most significant outcomes from the march was when one of the students I mentor from Richmond, California (one the most disadvantaged communities in America and my hometown) told me that **he and his family participated after finding out I was an organizer —** because seeing fellow young people of color engage in activism is uplifting.

– *Robin D. López*

SAN FRANCISCO
CALIFORNIA

The more time I spend being a physicist, the stranger it feels every time someone tells me

"YOU DON'T LOOK LIKE A PHYSICIST."

Years ago, my initial response was to get flustered, and ask "why is that hard to believe?"

It upset me. The responses would range from "you look more like you'd be in political science" to "you can't be a physicist, you must be kidding" in exasperated tones. The words I spoke didn't match my appearance — I'm female, like to wear makeup, and look young for my age. None of those attributes fit the traditional image of a physicist.

Honestly, I find the shock disappointing. I have worked tirelessly to earn my PhD in Physics at Harvard University, and it felt like a blow to the stomach every time anyone questioned my "fit." It felt like they were questioning my life's passion.

The truth is, the word "physicist" conjures up an accurate and discouraging image of the current state of affairs: recent reports suggest that only ~16% of physics faculty members are women, while the rest are male, and mostly white.

People have developed a strong association between the way Albert Einstein and other famous scientists look, and what physicists are supposed to look like. Not just him, but all other rock star physicists (except Marie Curie and Maria Goeppert-Mayer, the only two women to ever win Nobel prizes in Physics out of a total of 200) are/were white and male: Feynman, Heisenberg, Schroedinger, Maxwell, Dirac, Planck, etc.

We see stereotypical images of physicists in our everyday lives that get etched into our minds, and we gravitate towards media that reinforces these stereotypes. Everything eventually leads to a set of implicit biases in our minds.

When people are surprised to meet physicists who look different, it isn't really their fault. It may be the first time they've met a physicist, or the first time they've seen one that looks so starkly different than Einstein or Sheldon Cooper from the Big Bang theory. It just feels weird.

These stereotypes exist across the board in STEM, in many fields ranging from computer science to genetics. Science is hard enough. I know that for many scientists that look like me, it can be even more challenging, and we are at risk of losing them from science forever.

This realization compelled me to take action and co-develop the "I Am a Scientist" Campaign. It's a story-based initiative focused on increasing and embracing diversity in STEM. We believe it is crucial to start with changing the image that students (and teachers) have of what it means to be a scientist. We believe diversifying this image includes and extends beyond categories such as race, gender, sexuality, physical ability, and religion. The campaign aims to also share the person behind the science — pet passions, unexpected hobbies, challenges, interesting trivia — so that middle and high school students can see themselves in the stories.

The campaign is a collaboration between artists, educators, and researchers in the Boston area. It is being organized by Covi Education, a nonprofit dedicated to improving the relationship between science and society.

Here's my small plea: next time you meet a scientist that doesn't look like you thought they would, try your hardest to contain your look of surprise. They are at the forefront of redefining who you think can be a scientist, and you are an important part of that journey.

May the Force be with you.

Nabiha Saklayen, PhD
CEO and Co-Founder of Cellino Biotech, Inc.
Creator, "I Am a Scientist" Campaign

SCIENCE HAS THE POWER TO HELP US FIND REAL ANSWERS TO THE ISSUES OF EDUCATION, HEALTH, URBAN VIOLENCE, AND SUSTAINABLE DEVELOPMENT IN BRAZIL. BUT SCIENCE NEEDS OUR HELP IN TELLING BRAZILIANS ABOUT IT. WE MARCHED TO RAISE AWARENESS.

– Natalie Cella
Organizer

SÃO PAULO
BRAZIL

FACT-RESISTANT HUMANS ENDANGER ALL SPECIES

BATON ROUGE, LA, USA

I come from a long line of beekeepers (I'm fourth generation!) and **I am concerned that without adequate support for the EPA and USDA, the health of the honeybee could decline even further.**

– Taylor Reams
Greensboro, NC, USA

I'm motivated to bring awareness to people and organizations about the importance of multiple nations working together to save the environment.

I march for the parts of the world where there is no funding, where people have to make do with what they have to continue the fight against loss of diversity and climate change.

– Preston Marshall Thompson
Bowling Green, OH, USA

SUPPRESSION OF SCIENCE IS A REAL BUZZ KILL

PORTLAND, OR, USA

I'M NOT A SCIENTIST.
I DON'T HAVE A FORMAL SCIENCE EDUCATION. WHAT I DO HAVE IS A PASSION FOR SCIENCE, MEDICAL SCIENCE SPECIFICALLY.
IT'S FASCINATING!

I co-organized the March for Science – Houston because science plays a key role in my life. I am a medical mystery. From a rare type of Diabetes to Mitochondrial Disorder and an unknown eye condition, I have spent my entire life attempting to solve it.

Thirty years later, even at some of the best medical institutions in the country, it has been impossible to decipher the puzzle. My case is so complex and rare that I have been accepted to the Undiagnosed Disease Network, a research program at Baylor College of Medicine in Houston funded by the NIH.

Along the way I became afflicted with severe, debilitating pain on my right leg, often requiring mobility aids. For ten years it remained undiagnosed, until 2016, when I met Dr. Brian Bruel, Interventional Pain Medicine Physician, Director of Pain Medicine at Baylor College of Medicine and Associate Professor of Physical Medicine and Rehabilitation at Baylor College of Medicine. When I first entered his office, I felt as if I was meeting an old friend. His kindness and wisdom are conveyed in his work and the way he teaches the residents and students at his side.

Dr. Bruel immediately had a diagnosis, which he confirmed following diagnostic testing – I had Complex Regional Pain Syndrome (CRPS), possibly the most painful condition known to medical science. Treatment is experimental, but when I informed Dr. Bruel about the March for Science in Houston, he was determined to ensure I could march on my own two feet.

Not only did he accomplish his goal, but he marched right alongside me!

Alejandra Ruley
Director, March for Science – Houston

Proud LATINA marching for SCIENCE

PHILADELPHIA, PA, USA

Just a Black Woman marching for science in SF.

Too bad that deserves a picture.

SAN FRANCISCO, CA, USA

SCIENCE Making the world more FABULOUS for 4.5 billion years!

MADISON, WI, USA

PROUD TRANS SCIENTIST

TALLAHASSEE, FL, USA

TWIN CITIES, MN, USA

While I was a part of a much larger collective during the Science March,

I march everyday within my field to push it to be more inclusive from within.

As a queer, multiracial (Chinese, Scots-Irish, Welsh, and Cherokee) female scientist with dyslexia, I believe our way to collectively improve science is through community engagement. Though I sometimes struggle within science to carve a place out for myself, being an earth scientist is an important part of my identity. **I believe that thinking differently has allowed me to be open to different opinions and to grow as a person.**

I march for science inside the lab, the classroom hallways, and in the streets for a more inclusive field of science. Not just for various labels or categories of diversity, but for diversity of thought within and outside STEM. We need to listen to each other in order to move forward and make groundbreaking discoveries. Especially when we disagree.

– Sami
Palo Alto, CA, USA

The first time my daughter told me she wanted to be a scientist, I think she was two. I did a funny thing then, *and believed her.*

I signed her up for classes at the zoo, the nature center, farm trips, science labs, and later lemur sanctuaries, the Monterey Bay Aquarium, and rehabilitation facilities across the US. We got to be a part of the mission to help the ducklings of the US National Reflecting Pools get ramp access to increase their survival rates. These are our vacations.

She is only nine now, but science is part of her identity. In school, she spent her playground minutes giving mini speeches about earthworms and convincing classmates not to deliberately step on them. She even had her little brother watching his step!

I chose to homeschool her so she could add courses like forensics and anthropology to her lessons. Our little community offers a lot of wonderful opportunities to engage in the sciences, but it isn't enough. There is a struggle to fill the classes she enjoys most. One even became known as the "Emma and Nathan" program, because my children were often the only ones there. The last time I got a call that a class was canceled due to low enrollment, I said, "Please wait, give me a few hours and let me see if I can get the spots filled." I begged every homeschool community I knew, and we did get the classes filled. It felt amazing to see how a little advocacy saved those excellent programs.

That's why, when I heard none of the larger groups in our area were able to plan a March, I put out messages in community groups looking for individuals to start a grassroots event. A handful of passionate women did the same, and we found each other. A zookeeper, a naturalist, an advocate, a student, a journalist and me, a mom.

None of us was at the top of our field, or had great connections, but we had excellent chemistry and the planning pieces fell perfectly into place more often than not. We raised $1,400 in science scholarships, which was our way of turning our message into action. And my daughter and son were there next to me on the stage when I got to thank everyone for their donations.

BUT REALLY, I WAS ONLY UP THERE BECAUSE A LITTLE GIRL WANTED MORE, AND **I WILL NOT STAND FOR ANYONE GETTING IN HER WAY.**

Megan Pulley
Educator

BERLIN
GERMANY

We marched for science and society. We want critical thinking to take back its place in the democratic process. Scientists cannot keep silent. When we hear alternative historical facts, we, as Germans, are particularly alarmed.

Our march sang a very old German folk song called "Die Gedanken Sind Frei," written in 1842 by Hoffmann von Fallersleben. The words translate to: "The thoughts are free, they remain free and they cannot be controlled by anyone." The song was the lead song in the German Revolution of 1848 when court and church suppressed the rebellion of the citizens. We sang it as a signal of support for suppressed science in Turkey, Hungary, Poland, and wherever in the world science is not free.

– *Franz Ossing*
Organizer

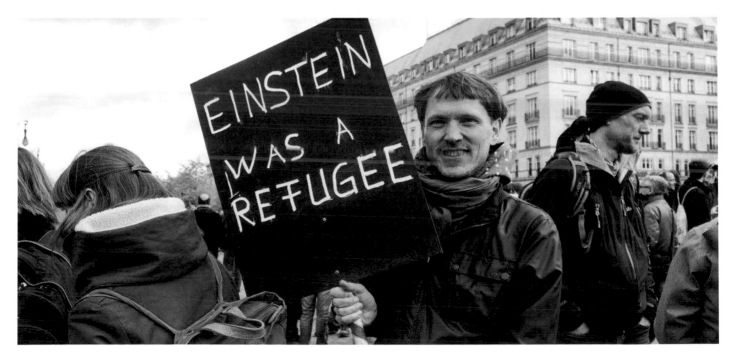

WHAT DO WE WANT?
Evidence Based Science
WHEN DO WE WANT IT?
After Peer Review

SCIENCE

SILICON VALLEY, CA, USA

I TELL STUDENTS SCIENCE CAN CHANGE THE WORLD DON'T MAKE ME A LIAR

WASHINGTON, DC, USA

LOS ANGELES
MARCH FOR SCIENCE
EARTH DAY

SPONSORED BY: NEXTGEN CLIMATE AMERICA   SUPPORTED BY: MARCH   FlashMarketing   ZeusVision   see   Nerdist   LA MAKER SPACE

WHEN I WAS YOUNG, I REMEMBER LOOKING THROUGH MY BEDROOM WINDOW, THINKING ABOUT ALL THE PEOPLE IN THE WORLD

LIVING UNDER ONE SKY.

In elementary school, I went on several field trips to the science museum, did chemistry experiments at home, and spent weekends at math competitions. These early experiences set me on a path toward a career in math and science.

At a recent dinner under space shuttle Endeavor, I celebrated with Dr. Mae Jemison for the 25th anniversary of her historic spaceflight. My path from looking up at the sky, to sitting with an awe-inspiring astronaut required persistence and acting on opportunities. "Seizing opportunity" is part of the story we don't share with young people often enough. Society tells us that the most successful among us worked really hard, but opportunities often disguise themselves as hard work. Having a successful career requires the courage to say yes to opportunities, even those that make you uncomfortable.

I've earned my BS in computer science and my MS and PhD in neuroscience. I've taught students from elementary school through graduate school, helped develop government policies at the California state capitol and served on nonprofit boards. I've met Nobel prize winners, elected officials, and university presidents. I've given talks to audiences with tens of thousands of people and mentored one-on-one. Opportunities like these and my liberal arts education led me to a career in service to the community. But it's fundamentally my family who inspired me.

I live at the intersection of race, gender, and class as a mixed-race woman, who grew up on welfare, raised by my dad after my parents divorced. Through many difficulties, we created our family: my dad, my brother, and me. I had early opportunities to learn from my father. He taught us about nature and space on nightly walks. He taught us to analyze noises to fix our broken car. And he read to us every night until middle school.

Growing up poor in a single parent home is not unique, but it's not a common background to find in my career field. People tend to solve the problems they themselves have experienced. And, it shouldn't be a surprise that I tackle challenges facing low-income or marginalized communities. I build diverse teams, speak with community members, gather evidence, and seek innovative solutions to improve our world.

We need to inspire kids to use the scientific method as a tool to re-create the world into one that they want to live in — and give them the opportunities they need to become successful.

We need to teach people how to become critical thinkers and problem solvers using observation and analysis to understand their struggles. We need communities to understand how innovations would benefit them. And we need to choose leaders that recognize science and use it as a tool for social good.

I've had many jobs, but only one career: using science to improve the lives of others living under the same great sky. Science has the power to make the world a better place.

Tepring Piquado, PhD
Neuroscientist
Life Member, SACNAS

SAN DIEGO, CA, USA

KANSAS CITY, MO, USA

LONDON, UK

A big part of the reason I march is for my Uncle Jamie.

Upon returning from Vietnam, he became an engineer. He is currently suffering from Alzheimer's and will be the third family member my husband and I have seen succumb to the disease in the past four years.

I march because we need an administration open to funding research, medical advancements, and protecting the environment for future generations.

We need an administration that favors science over unproven opinions. PERIOD.

I march for one of my childhood heroes, who I know would be marching with me if he could.

– Jeanette Marlene
Liverpool, NY, USA

JUNEAU, AK, USA

I AM AN UNDOCUMENTED STUDENT WHOSE
LIFE HAS BEEN CHANGED BY SCIENCE.

I was one of two kids, born into poverty, and by the age of 6 I was working full time in construction just trying to survive. We were born to fail, but we risked everything to change our lives. If you had asked me the day I was crossing the border at 13, if could I see myself graduating from college, I would have said never! Not in my wildest dreams.

In 2015, I received my chemistry degree from University of California, Irvine, and will become a future medical doctor.

I work for a nonprofit that promotes science all around the nation for underserved and underrepresented communities. After the March of Science, I was even more motivated to organize undocumented STEM students, so with the help of many, I organized the first annual UndocuSTEM Conference.

Please do not forget that minorities, like me who are vulnerable and under attack, need you to stand up for us as we stand up for ourselves. I encourage you all to extend your help and to make sure minority students in science become scientists.

 Rene Amel Peralta
Education Programs Coordinator, Great Minds in STEM
Co-Founder, UndocuSTEM

TWIN CITIES, MN, USA

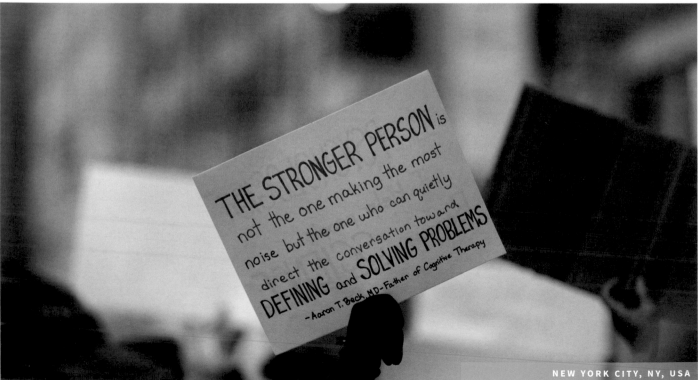

THE STRONGER PERSON is not the one making the most noise but the one who can quietly direct the conversation toward DEFINING and SOLVING PROBLEMS
- Aaron T. Beck, MD-Father of Cognitive Therapy

NEW YORK CITY, NY, USA

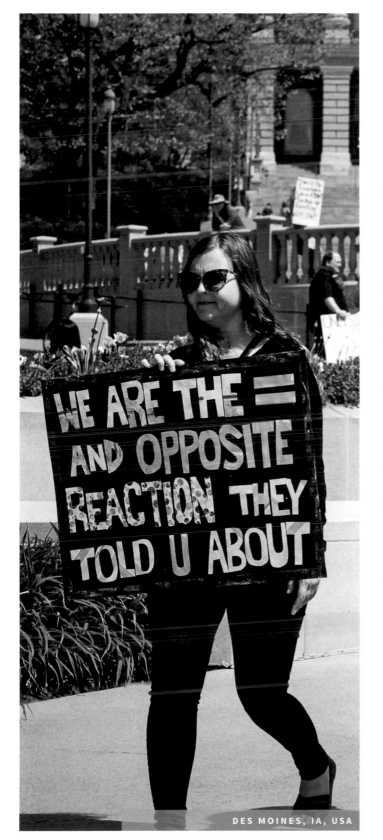

DES MOINES, IA, USA

I WANT MY SON TO GROW UP IN

A WORLD BASED ON FACTS, NOT FEAR...

...a world that celebrates diversity and lifts people up. I became a scientist because I liked the idea of using data and facts to solve problems. But I left academia for the messy not-always-fact-based world of science policy because I wanted to help solve the pressing problems there. As a proud scientist at the Union of Concerned Scientists, I'm amazed now to see what this movement for science has become and I have hope that my son can have a fact-based future.

–Gretchen Goldman, PhD
Washington, DC, USA

PORTLAND, OR, USA

SAN FRANCISCO, CA, USA

TWENTY-THREE YEARS AGO, SCIENCE WASN'T ABLE TO SAVE MY PREMATURE BABY'S LIFE. TEN YEARS AGO, ADVANCEMENTS IN SCIENCE SAVED MY 2ND PREMATURE BABY'S LIFE. For my angel and my ten year old, I March.

– *Jobyna McCarthy, Missoula, MT, USA*

ATLANTA, GA, USA

KANSAS CITY, MO, USA

RALEIGH, NC, USA

TWIN CITIES, MN, USA

WASHINGTON, DC, USA

I like science because I like making things. It's like you're just playing, but then you have something cool to show your mom at the end.

– Bella, Age 5
Portland, OR, USA

AMHERST, MA, USA

WASHINGTON, DC, USA

SCIENCE SERVING THE COMMON GOOD

SCIENCE PURSUING TRUTH SAVING THE WORLD

SCIENCE PROTECTING OUR COMMUNITIES

SCIENCE SPEAKING TRUTH TO POWER

Our team at The Natural History Museum helped to develop the mission statement and overall messaging strategy for the march — moving beyond science boosterism to uplift science that serves the common good, speaks truth to power, and protects the people and places we love. We designed signs for the front of the march, worked with our scientist collaborators to place op-eds, and facilitated feature segments on several news shows. We're proud to have helped recruit several science museums and institutions to step out of their comfort zones to endorse the march and play a more active role in championing a safe and equitable future for all. Finally, we collaborated with the Center for Native Peoples and the Environment to organize a letter affirming the role of Indigenous Science and traditional ways of knowing, signed by 1,700 Indigenous scientists, tribal leaders, and allies.

All told, the March for Science was unprecedented: never before have scientists and science institutions stepped into the public sphere in such a big way. There have historically been fierce debates around the question of neutrality in science. We've taken a quantum leap forward in normalizing science activism and injecting momentum and political potential into the single most trusted constituency in America. Scientists mobilized en masse to champion sacred notions of truth, defend climate science, EPA science, and health research, and advocate alongside traditionally marginalized communities hit hardest by the war on science.

THE MARCH FOR SCIENCE WASN'T A MARCH FOR SCIENTISTS — IT WAS A MARCH FOR ALL OF US.

Scientists are waking up to the need to advocate on behalf of their findings and the communities that are most impacted by air and water contamination, climate change, and other social and environmental concerns.

Beka Economopoulos
Executive Director, The Natural History Museum
New York City, NY, USA

THERE ARE COUNTLESS REASONS I'M MARCHING, BUT
**SAVING OUR PLANET AND ANIMALS FROM THE HARMFUL
EFFECTS OF CLIMATE CHANGE IS AT THE TOP OF THE LIST.**

I studied wildlife management in South Africa and wildlife conservation
in Thailand. After graduation I went on to become a giraffe keeper.

Taking care of giraffes helped to give me the patience I need to raise
kids!

 Kate Cortelyou
Mom
Former Giraffe Keeper

"SWIM FOR SCIENCE"

I MARCH FOR THIS PLANET THAT HAS GIVEN ME EVERYTHING...

YOSEMITE NATIONAL PARK

YELLOWSTONE NATIONAL PARK

MARCH FOR SCIENCE
APRIL 22, 2017
NORTH POLE

NORTH POLE

BECAUSE I OWE HER ONE.

– Amanda Sargent
Boston, MA, USA

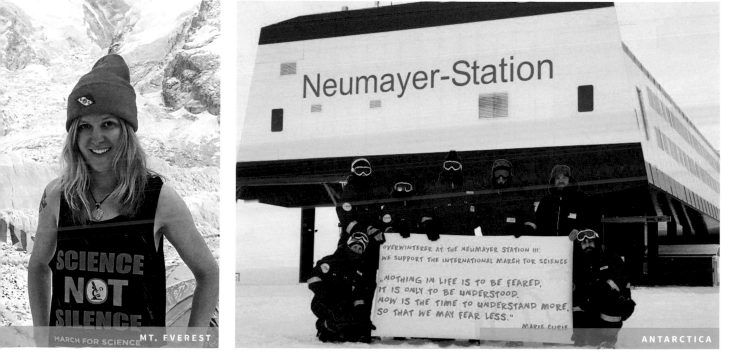

SCIENCE NOT SILENCE

MARCH FOR SCIENCE   MT. EVEREST

Neumayer-Station

OVERWINTERER AT THE NEUMAYER STATION III:
WE SUPPORT THE INTERNATIONAL MARCH FOR SCIENCE

"NOTHING IN LIFE IS TO BE FEARED,
IT IS ONLY TO BE UNDERSTOOD.
NOW IS THE TIME TO UNDERSTAND MORE,
SO THAT WE MAY FEAR LESS."

MARIE CURIE

ANTARCTICA

Rain, cold, and allergies couldn't keep my ten-year-old and me from celebrating science!

My son can be shy at times. He was so excited about participating in the March for Science that he spoke to complete strangers he came across all day, asking what inspired them to pursue science when they were kids, and sharing his own interest in becoming a marine biologist.

As the son of Muslim immigrants, he already understands the importance of representation and making America, and the world, a wonderful place to live in through the wonders of science.

– *Anil Ahmed*
Ashburn, VA, USA

SAVANNAH, GA, USA

E = mc SAD

SEATTLE, WA, USA

THEY TOLD ME TO BRING A SINE

MISSOULA, MT, USA

IT'S SO BAD... EVEN THE INTROVERTS CAME OUT!

COLUMBUS, OH, USA

THIS IS NOT NORMAL

$Y = \chi^2(x)$ $df = 3$

NEW YORK CITY, NY, USA

When I was in the Bush Administration and most recently at the White House in the Obama Administration, I had the opportunity to hear the stories of amazing Americans from all over the country. One of my favorite things was when we turned the White House into a science fair. We invited kids from all around the country to show off their research. And wow.

Take for example, Nathan, who was 15 at the time and built machine learning and artificial intelligence algorithms to detect genetic mutations that are likely to cause cancer. Simon-Peter, Maya, and Grayson — at 13 and 14 — designed a new prosthetic leg that will allow an amputee to hike, manage uneven terrain, and even skateboard! And 17-year-old Olivia developed a rapid, portable, and inexpensive diagnostic test for detecting the Ebola virus. These kids are from every walk of life. From every part of America. Spend five minutes with them and you'll leave the conversation with an incredible inferiority complex.

Science and engineering can't wait. Slowing down isn't an option. Kids in Flint, Michigan, still don't have clean water. The climate is changing. The next pandemic could be around the corner. We have a growing world population that needs to be fed and educated. Our national defenses will continue to depend on innovation. And cancer and rare diseases continue to take too many lives.

When I was in the White House, I also met Jennifer Bittner, who wrote to the president — yes we did read the letters you sent, and actually respond. Here's the first thing you need to know about Jennifer: she's a beautiful person, a wonderful wife to her husband Rod, and a phenomenal mother to their son and the child that is on the way. She is a force of nature. This is what Jennifer said in her own words:

I was diagnosed with Stage IV metastatic breast cancer that had spread to my liver, lungs, adrenal gland, spleen, ovaries, spine, hips, ribs, femur, scapula, clavicle, and many other bones. One doctor said I'd be in hospice within three months. While it's hard to put into words how grateful I am to have had every single moment of this wonderful life the past few years, it's certainly not easy. I've been on chemotherapy or other treatment every single day. I've endured multiple surgeries and radiation, as well as the anxiety of weekly tumor markers and scans every three months. My treatment side effects are challenging and at times, debilitating. Despite all of this, the average life expectancy of someone with metastatic breast cancer is just three short years.

Three years is far too short when you're not even halfway through life. Research is absolutely critical to extending and improving the lives of people with cancer — especially now that we are on the precipice of some incredible medical breakthroughs such as immunotherapies. That's why it's so critical that research be fully funded. We are so close but most of us can't wait much longer.

> ## Cancer doesn't wait. Rare diseases don't wait. Pandemics don't wait. Our kids can't wait.

When we come together as a nation; when we focus as a community; and when we are relentless in our determination to support all the sciences, we will keep science in its rightful place leading our country forward. Forward for a positive future for my children, your children, the world's children, and our children's children.

Science can't wait. Let's get to work.

DJ Patil, PhD
Former U.S. Chief Data Scientist

ALABAMA AUBURN **BIRMINGHAM** HUNTSVILLE **MOBILE** MONTGOMERY **ALASKA** ANCHORAGE **DILLINGHAM** FAIRBANKS **HOMER** JUNEAU **KETCHIKAN** PALMER **SEWARD** SKAGWAY **TOK** TOOLIK LAKE **ARIZONA** CHINO VALLEY **FLAGSTAFF** LAKE HAVASU CITY **PHOENIX** SEDONA **TUCSON** WHITE MOUNTAINS **ARKANSAS** FAYETTEVILLE **FORT SMITH** LITTLE ROCK **CALIFORNIA** BERKELEY **CAMBRIA** CHICO **COACHELLA VALLEY MUSIC AND ARTS FESTIVAL** FORT BRAGG **FRESNO** FULLERTON **HANFORD** HAYWARD **HEMET** HUMBOLDT **KELSO DEPOT** LAKE TAHOE **LIVERMORE** LONG BEACH **LOS ANGELES** MERCED **MODESTO** MONTEREY **NEVADA COUNTY** OJAI **OLYMPIC VALLEY** PACIFICA **PALM SPRINGS** PALMDALE **PASADENA** QUINCY **REDDING** RIDGECREST **RIVERSIDE** SACRAMENTO **SAN DIEGO** SAN FRANCISCO **SAN JOSE** SAN LUIS OBISPO **SANTA BARBARA** SANTA CRUZ **SANTA ROSA** SONORA **STOCKTON** TEHACHAPI **WALNUT CREEK** YOSEMITE **YOSEMITE VALLEY** **COLORADO** ASPEN **AVON** BRECKENRIDGE **CARBONDALE** COLORADO SPRINGS **DENVER** ESTES PARK **FORT COLLINS** GRAND JUNCTION **GUNNISON** LA JUNTA **TELLURIDE** **CONNECTICUT** EAST **HADDAM** EAST LYME **HARTFORD** NEW HAVEN **WASHINGTON D.C.** **DELAWARE** LEWES **NEWARK** **FLORIDA** CLEARWATER **FORT LAUDERDALE** FORT PIERCE **FORT WALTON BEACH** GAINESVILLE **HUDSON** JACKSONVILLE **LAKELAND** MIAMI **NAPLES** NEW SMYRNA BEACH **ORLANDO** PALM BEACH COUNTY **PANAMA CITY** PENSACOLA **SARASOTA** SPACE COAST **ST. AUGUSTINE** ST. PETERSBURG **TALLAHASSEE** WEST PALM BEACH **GEORGIA** ATHENS **ATLANTA** AUGUSTA **BRUNSWICK** SAVANNAH **STATESBORO** VALDOSTA **HAWAII** HILO **HONOLULU** LIHUE **MAUI** **IDAHO** BOISE **IDAHO FALLS** MOSCOW **POCATELLO** **ILLINOIS** CARBONDALE **CHAMPAIGN** CHARLESTON **CHICAGO** GENEVA **NORMAL** PALATINE **PEORIA** ROCKFORD **SPRINGFIELD** **INDIANA** EVANSVILLE **INDIANAPOLIS** LAFAYETTE **SOUTH BEND** TERRE HAUTE **IOWA** CEDAR FALLS **DAVENPORT** DECORAH **DES MOINES** INDEPENDENCE **IOWA CITY** TIPTON **KANSAS** MANHATTAN **TOPEKA** WICHITA **KENTUCKY** BOWLING GREEN **LEXINGTON** LOUISVILLE **PADUCAH** **LOUISIANA** BATON ROUGE **LAFAYETTE** MONROE **NEW ORLEANS** SHREVEPORT **MAINE** GOULDSBORO **MACHIAS** ORONO **PORTLAND** SANFORD **UNITY** **MARYLAND** ANNAPOLIS **OCEAN CITY** **MASSACHUSETTS** AMHERST **BOSTON** FALMOUTH **GREAT BARRINGTON** MANSFIELD **PITTSFIELD** WORCESTER **MICHIGAN** ALBION **ALPENA** ANN ARBOR **BIG RAPIDS** CHEBOYGAN **DETROIT** GRAND RAPIDS **HOUGHTON** KALAMAZOO **LANSING** MARQUETTE **MIDLAND** PETOSKEY **SAULT** STE. MARIE **YPSILANTI** **MINNESOTA** ALEXANDRIA **BEMIDJI** BRAINERD **DULUTH** GRAND MARAIS **GRAND RAPIDS** MOORHEAD **MORRIS** NEW ULM **NORTHFIELD** PARK RAPIDS **ROCHESTER** ST. PAUL **MISSISSIPPI** HATTIESBURG **LONG BEACH** OXFORD **MISSOURI** COLUMBIA **JOPLIN** KANSAS CITY **MARYVILLE** ROLLA **SPRINGFIELD** ST. JOSEPH **ST. LOUIS** **MONTANA** BILLINGS **BOZEMAN** GREAT FALLS **HELENA** MISSOULA **NEBRASKA** HASTINGS **KEARNEY** LINCOLN **OMAHA** **NEVADA** LAS VEGAS **RENO** SPRING CREEK **NEW HAMPSHIRE** CONCORD **PORTSMOUTH** **NEW JERSEY** ATLANTIC CITY **PRINCETON** TRENTON **NEW MEXICO** ALBUQUERQUE **LAS CRUCES** SANTA FE **SILVER CITY** SOCORRO **TAOS** **NEW YORK** ALBANY **BINGHAMTON** BUFFALO **CORNING** EAST MEADOW **ITHACA** PLATTSBURGH **POUGHKEEPSIE** ROCHESTER **ROCKVILLE CENTER** **SARATOGA SPRINGS** SCHOHARIE **STONY BROOK** SYRACUSE **UTICA** WATERTOWN **WELLSVILLE** **NORTH CAROLINA** ASHEVILLE **BEAUFORT** CHARLOTTE **ELIZABETH CITY** GREENSBORO **MORGANTON** RALEIGH **WASHINGTON** WILMINGTON **NORTH DAKOTA** FARGO **GRAND FORKS** **OHIO** ATHENS **CINCINNATI** CLEVELAND **COLUMBUS** DAYTON **DELAWARE** FINLEY **MANSFIELD** MOUNT VERNON **OXFORD** TOLEDO **WOOSTER** YELLOW SPRINGS **YOUNGSTOWN** ZANESVILLE **OKLAHOMA** OKLAHOMA CITY **TULSA** **OREGON** ASHLAND **BEND** COOS BAY **CORVALLIS** EUGENE **GRANTS PASS** KLAMATH FALLS **NEWPORT** PENDLETON **PORTLAND** ROSEBURG **SALEM** SISTERS **ST HELENS** **PENNSYLVANIA** BEAVER **BETHLEHEM** BRADFORD **DOYLESTOWN** EAST STROUDSBURG **ERIE** HAWLEY **LANCASTER** MEADVILLE **PHILADELPHIA** PITTSBURGH **SELINSGROVE** SHARON **STATE COLLEGE** **PUERTO RICO** SAN JUAN **RHODE ISLAND** PROVIDENCE **SOUTH CAROLINA** CHARLESTON **CLEMSON** COLUMBIA **GREENVILLE** MYRTLE BEACH **SPARTANBURG** AIKEN **SOUTH DAKOTA** ABERDEEN **PIERRE** RAPID CITY **SIOUX FALLS** **TENNESSEE** CHATTANOOGA **KNOXVILLE** MEMPHIS **MEMPHIS RALLY** NASHVILLE **TEXAS** ALPINE **AMARILLO** AUSTIN **BEAUMONT** COLLEGE-STATION **CORPUS CHRISTI** DALLAS **DENTON** EL PASO **FORT WORTH** GEORGETOWN **HOUSTON** LUBBOCK **LUFKIN** MIDLAND **NEW BRAUNFELS** SAN ANTONIO SHERMAN **WICHITA FALLS** WIMBERELY **US VIRGIN ISLANDS** CHARLOTTE AMALIE **UTAH** CEDAR CITY **LOGAN** MOAB **PARK CITY** SALT LAKE CITY **SPRINGDALE** ST GEORGE **VERMONT** BRATTLEBORO **BURLINGTON** MONTPELIER **RUTLAND** **VIRGINIA** BLACKSBURG **CHARLOTTESVILLE** LYNCHBURG **MARTINSVILLE** NORFOLK **RICHMOND** STAUNTON **WILLIAMSBURG** WINCHESTER **WASHINGTON** BELLINGHAM **CHEHALIS** COUPEVILLE **ELLENSBURG** KENNEWICK **OLYMPIA** PORT ANGELES **PULLMAN** SEATTLE **SHELTON** SPOKANE **TACOMA** WENATCHEE **WHIDBEY** WHITE SALMON **YAKIMA** **WEST VIRGINIA** BUCKHANNON **CHARLESTON** HUNTINGTON **MORGANTOWN** **WISCONSIN** APPLETON **ASHLAND** EAU CLAIRE **GREEN BAY** KENOSHA **LA CROSSE** MADISON **MARSHFIELD** MILWAUKEE **MINOCQUA** OSHKOSH **RICE LAKE** STEVENS POINT **WEBSTER** **WYOMING** CODY **JACKSON** LARAMIE **OLD FAITHFUL, YELLOWSTONE NATIONAL PARK** PINEDALE

ARGENTINA CIUDAD AUTÓNOMA DE BUENOS AIRES SAN MIGUEL DE TUCUMAN, TUCUMAN AUSTRALIA BRISBANE CAIRNS CANBERRA HOBART MELBOURNE PERTH PORT MACQUARIE SYDNEY TOWNSVILLE ADELAIDE, SOUTH AUSTRALIA LAUNCESTON, TAS AUSTRIA VIENNA BAHAMAS HOPE TOWN, ABACO BANGLADESH DHAKA BELGIUM BRUSSELS BHUTAN WANGDUE, WANGDUE PHODRANG BOSNIA AND HERZEGOVINA SARAJEVO BRAZIL BRASILIA FLORIANÓPOLIS MANAUS PORTO ALEGRE RIO DE JANEIRO SÃO CARLOS GOIÂNIA, GOIÁS BELO HORIZONTE, MINAS GERAIS ITAJUBÁ, MINAS GERAIS CURITIBA, PARANÁ PATO BRANCO, PARANÁ BELÉM, PARÁ PETROLINA, PERNAMBUCO RECIFE, PERNAMBUCO NATAL, RIO GRANDE DO NORTE SÃO PAULO, SÃO PAULO BRITISH VIRGIN ISLANDS CANE GARDEN BAY BULGARIA SOFIA CANADA CALGARY HAMILTON MONTREAL OTTAWA PRINCE GEORGE TORONTO VANCOUVER VICTORIA EDMONTON, ALBERTA LETHBRIDGE, ALBERTA WINNIPEG, MANITOBA ST. JOHN'S, NEWFOUNDLAND HALIFAX, NOVA SCOTIA KITCHENER-WATERLOO, ONTARIO LONDON, ONTARIO SUDBURY, ONTARIO WINDSOR, ONTARIO SASKATOON, SASKATCHEWAN YARMOUTH, NOVA SCOTIA CAYMAN ISLAND LITTLE CAYMAN GEORGE TOWN, GRAND CAYMAN CHILE ANTOFAGASTA SANTIAGO VIÑA DEL MAR CONCEPCION, REGION DEL BIO BIO CHINA HONG KONG COLOMBIA MEDELLIN, ANTIOQUIA BOGOTÁ, CUNDINAMARCA BUCARAMANGA, SANTANDER CALI, VALLE DEL CAUCA COSTA RICA SAN PEDRO, SAN JOSÉ CROATIA SPLIT ZAGREB RIJEKA, CROATIA CZECH REPUBLIC BRNO, CZECH REPUBLIC PRAGUE, CZECH REPUBLIC DENMARK COPENHAGEN AARHUS, JYLLAND ECUADOR URCUQUI, IMBABURA ESTONIA TALLINN FEDERATED STATES OF MICRONESIA KOLONIA POHNPEI FINLAND HELSINKI FRANCE CLERMON-FERRAND GRENOBLE LILLE LYON MONTPELLIER NANCY NANTES NICE PARIS STRASBOURG TOULOUSE MARSEILLE, BOUCHES-DU-RHONE THONON-LES-BAINS, HAUTE-SAVOIE BORDEAUX, NOUVELLE AQUITAINE CIVRAY, VIENNE GERMANY BERLIN BONN COLOGNE DRESDEN FRANKFURT FREIBURG HAMBURG HEIDELBERG KOBLENZ LEIPZIG MUNICH MÜNSTER STUTTGART TUEBINGEN, BADEN-WUERTTEMBERG GREIFSWALD, MECKLENBURG-VORPOMMERN ROSTOCK, MECKLENBURG-VORPOMMERN GOETTINGEN, NIEDERSACHSEN HELGOLAND, SCHLESWIG-HOLSTEIN JENA, THÜRINGEN GHANA ACCRA GREENLAND KANGERLUSSUAQ, QEQQATA GUAM AGANA, GU GUYANA GEORGETOWN HUNGARY BUDAPEST ICELAND REYKJAVÍK INDIA COIMBATORE, TAMIL NADU HYDERABAD, TELANGANA IRAQ SULAYMANIYAH, AS-SULAYMANIYAH ITALY CASERTA FLORENCE MILAN NAPOLI PALERMO POTENZA RIMINI ROME JAPAN TOKYO TSUKUBA TSUKUBA IBARAKI LATVIA RIGA, LATVIA LITHUANIA VILNIUS LUXEMBOURG LUXEMBOURG CITY, LUXEMBOURG MALAWI BLANTYRE MEXICO MEXICO CITY SAN LUIS POTOSÍ ENSENADA, BAJA CALIFORNIA IRAPUATO, GUANAJUATO GUADALAJARA, JALISCO MORELIA, MICHOACAN CUERNAVACA, MORELOS PUEBLA, PUEBLA VILLAHERMOSA, TABASCO XALAPA, VERACRUZ NEPAL LUKLA, SOLOKUMBU NETHERLANDS AMSTERDAM MAASTRICHT, LIMBURG NEW ZEALAND AUCKLAND CHRISTCHURCH WELLINGTON PALMERSTON NORTH, MANAWATU DUNEDIN, OTAGO NIGERIA FEDERAL CAPITAL TERRITORY, ABUJA NORWAY BERGEN BODØ OSLO TROMSØ TRONDHEIM STAVANGER, ROGALAND LONGYEARBYEN, SVALBARD NY-ÅLESUND, SVALBARD PANAMA PANAMA CITY PHILIPPINES QUEZON CITY POLAND WARSAW, MAZOWSZA PORTUGAL LISBOA REPUBLIC OF IRELAND DUBLIN ROMANIA BUCHAREST CLUJ-NAPOCA RUSSIAN FEDERATION KAZAN, TATARSTAN SERBIA BELGRADE, SERBIA SLOVAKIA BRATISLAVA SLOVENIA LJUBLJANA SOUTH AFRICA DURBAN CAPE TOWN, WESTERN CAPE SOUTH KOREA BUSAN SEOUL SPAIN GIRONA MADRID SEVILLE VALLADOLID, CASTILLA Y LEÓN BARCELONA, CATALONIA SWEDEN LULEÅ STOCKHLOM UPPSALA, UPPLAND UMEÅ, VÄSTERBOTTEN GOTHENBURG, WEST SWEDEN SWITZERLAND GENEVA TAIWAN TAIPEI TRINIDAD AND TOBAGO PORT OF SPAIN, TRINIDAD UGANDA KAMPALA, CENTRAL UKRAINE KYIV UNITED

600+ MARCHES
ONE GLOBAL MOVEMENT

The March for Science was a success because it was necessary — and with each passing day our continued advocacy is more and more crucial. Our movement is growing, and it needs your voice: together, we can advocate for science and protect the future of our communities.

It doesn't matter if you have three minutes or three months to devote to science advocacy: you can help our cause. Our movement needs scientists to share their work, educators to organize teach-ins and science fairs, artists to bring our stories to life, and community members to raise their voices. Consider your own skills and passion, and find a way to get involved.

April 22, 2017, was the first step.

NOW, WE KEEP MARCHING TOGETHER.

1. BUILD COMMUNITY.

Support for science comes from all communities — it cuts across professions and age groups, geographic locations and ethnicities. We are stronger, louder, and more effective when we join together. So get involved with your local science advocacy community. Many cities and towns already have promising local science initiatives that don't get the attention and support that they deserve — find and bolster them. Attend (or organize!) a science outreach event to learn about the work of scientists in your town and how it affects policy. Visit a local science museum, zoo, or aquarium. Reach out to student science groups or start a science book club. However you choose to engage, remember that building community relationships is the first step in safeguarding long-term science advocacy.

2. ADVOCATE FOR CHANGE WITHIN INSTITUTIONS.

Scientific institutions are the most prominent representatives of science in society. We need to shape them so that they support and engage in grassroots advocacy more than they have in the past. That means pushing our institutions to prioritize funding and reward advocacy and community projects. It means creating space for collective organizing at conferences. It means demanding our institutions stand up for underrepresented voices in science and address problems with access and diversity. It means supporting teachers and schools as they update and strengthen science curriculums. And it means asking our institutions to speak up, and not stay on the sidelines, when significant policy issues arise. If we want to create change and empower future generations of science communicators and activists, we must be thoughtful, vocal advocates, and actively push for our values within our institutions. After all, they are our institutions, and science should be steered by all of us.

3. VOTE FOR SCIENCE.

Marching was an inspiring start, but now we need to build political power for science. To make lasting change, we have to hold our leaders accountable to scientific evidence, and demand that they write and support good science policy. That means that we need to register eligible voters and organize them to show up at the polls — and encourage our friends and community members to do so too! It means calling our elected leaders and telling them to vote for science policies. If we want policymakers to take science seriously, we have to show them that we take voting seriously. But the pressure can't just begin and end around elections. Educate yourself on science policy and get in the habit of calling your local and national leaders, and showing up at town halls. Ask your representatives, and candidates striving to represent you in office, to create platforms that explicitly detail their position on science, and the role of evidence in policymaking. Your elected representatives are accountable to your voice — so use that voice to advocate for science!

Want more ideas? Visit **marchforscience.com** for more information on how to pursue science advocacy as an effective **21st-century champion for science**.

PHOTOGRAP

THANK YOU TO OUR CONTRIBUTORS FOR SHARING THEIR TALENT AND MEMORIES

HERS

PORTLAND, OR, USA

MONTEREY BAY AQUARIUM, USA

ACKNOWLEDGMENTS

The March for Science wishes to thank every single science advocate who attended or supported a march on April 22, 2017, and every volunteer who made sure there was somewhere to go. On all seven continents and in more than 600 locations around the world, we came together in unprecedented numbers. We showed the world that science will not be silenced, and that we can and will keep marching to champion scientific research, evidence-based policies, and an inclusive and equitable scientific community.

First and foremost, this is your story. Thank you for being part of it.

We would like to express our gratitude to our contributors — satellite teams, individuals, photographers — for sharing their memories, images, artwork, and words with us. Your stories help the world remember and understand why we came together, what we're fighting to build and protect, what we have left to learn, and the impressive range of ways that science can impact all of our lives.

We want to thank the team at MIT Press who believed in the importance of amplifying our message. Their belief that the March for Science Is not as just a moment in time — but rather an opportunity for lasting change — made this project possible. Thank you to Wilburforce Foundation and our many other generous donors for their time and support.

Finally, to the countless scientists and activists who invigorated our passion and inspired our ongoing work and to the diverse, dynamic group of satellite organizers, partner organizations, and community volunteers who continue to work together to secure the future of science advocacy: the March for Science movement exists because of you. We are humbled to have learned from and been strengthened by you, and hope that the March for Science earns its place as one piece in an enduring legacy of science advocacy.

 The March for Science Team